# Battleship Sailor

BY THEODORE C. MASON

NAVAL INSTITUTE PRESS
Annapolis, Maryland

Copyright © 1982
by the United States Naval Institute
Annapolis, Maryland

Third printing with corrections, 1988

Library of Congress Cataloging in Publication Data

Mason, Theodore, 1921-
    Battleship sailor.

    1. United States. Navy—Sea life. 2. Mason,
Theodore, 1921—    . 3. United States. Navy—
Biography. I. Title.
V736.M27          359'.0092'4 [B]          81–85440
ISBN 0–87021–095–5                          AACR2

Printed in the United States of America

To all my shipmates in the USS *California*
and especially to
Melvin Grant Johnson, RM3, USN,
and Thomas James Reeves, CRM, USN, Medal of Honor,
and Tom Gilbert, S1, USN,
who sleep forever at the
National Memorial Cemetery of the Pacific
in Honolulu

# Contents

# Foreword

In 1940, with the nation not yet fully recovered from the Great Depression, the prospects of going to college seemed bleak to Ted Mason. The job outlook was not much better, even for an honor grad of the local high school; nor was there much else of interest to look forward to. The Naval Reserve, however, offered the jobs and adventure that were so scarce in Placerville, California, and the lad volunteered for a year's active service as a radioman. His first duty station, after initial training, was the battleship *California* in Pearl Harbor, where he expected to be no more than twelve months. Afterwards he would return home far more employable, with interesting tales to tell. He did not expect to be in a war.

Mason, along with most of his buddies, was unaware of the ominous forces at work in the world and their possible impact on him. But understanding was dawning when the United States declared an emergency and extended all enlistments, his included, for the duration. And then everything became clear to everyone on 7 December 1941.

Through a twist of fate, courtesy of the USS *California*'s "watch, quarter, and station bill," Ted Mason's battle station on that shattered Sunday morning was in the open maintop of the proud battle-force flagship. Through another set of circumstances, there was nothing up there to fight with. Mason and his maintop mates, listening with avid, white-knuckled attention to their telephone headsets, could only be spectators at the most humiliating failure in American naval history. Until heavy ammunition boxes were painfully hauled up the mast there was nothing to do but curse the enemy, dodge fire from strafing aircraft, and watch the horrific attack.

Early on, they felt the *California* begin to heel over under them. With utter disbelief they watched the *Oklahoma* likewise begin to list—and then roll over and turn almost completely bottom up. They felt the incandescent heat and were racked by the insane roar when the *Arizona*'s forward magazines exploded, tore out the battleship's entire forward section, and peppered a half-mile radius with zinging chunks of hot debris.

Mason viewed this thunderous scene from a front-row seat. Despite the emotional involvement that this position afforded him, his description of the attack is highly disciplined, his prose always under control. Partly because of this his account is remarkably powerful. Although he writes from the perspective of forty years later, Mason skillfully maintains the point of view of his enlisted youth. A strict adherence to this outlook serves him particularly well in his convincing portrayal of enlisted life on the *California* during the months preceding Pearl Harbor. Mason recreates in microcosm the "battleship navy" of a time long ago, when shipboard ice cream was mediocre, cigarettes dirt cheap, and "short-arm" inspections routine. And he tells about his and his buddies' forays to familiar sailors' haunts—bars, bordellos, and YMCAs—and some not-so-familiar visits to churches and theaters.

Mason's account of these sometimes rambunctious liberties, salted with the small details that evoke time and place, give his book an authenticity lacking in a lot of popular wartime fiction and motion pictures. Along with the shrill call of the boatswain's pipe we hear the nostalgic strains of Glen Miller's "In the Mood"; from the stuffy confines of the main radio room, filled with the odor of sweat and the ubiquitous navy "joe," we are taken down a Seattle street glaring with Christmas advertisements that cater to a public just recently released from the tightest grip of depression. *Battleship Sailor* is, in fact, a sort of social history—fresh, entertaining, and honest. Ted Mason succeeds admirably in bringing to life these times, as they were, in peace and on the first day of war.

After that day, Mason felt fury. Blame for the disaster he deals out from the standpoint of a young sailor who had been conditioned to place unquestioning faith in the high-level planning that controlled his life, and who felt betrayed when this planning was demonstrated to be a complete fiasco. Many of Mason's closest acquaintances were killed deep below decks, including Chief Radioman Thomas J. Reeves, who died in the passageway alongside his submerged "main radio" while organizing an emergency ammunition train. Mason, as well as so many others at Pearl Harbor and elsewhere, attributed the demise of shipmates and friends, and the terrible damage inflicted on the great ships, to some profound shortcoming of the navy that suffered them to happen. Hurt and anger lend validity to his indictments.

In the vivid pages that follow, upon which Mason has set down his experiences of forty years ago, he performs for us a most valuable service by illuminating navy enlisted life, and an infamous day, in a way that allows us to see what we have never seen before. Ted Mason tells us what it was like, in those days, to be a battleship sailor.

EDWARD L. BEACH
*Captain, U.S. Navy (Retired)*

# Acknowledgments

Upon writing *finis* to a book project, an author can perhaps be forgiven his conceit in fancying himself an intrepid Captain James Cook, home safe from the sea after years of exploring uncharted waters and landing on hostile beaches. Only then, when riding to anchor, can he take time to drink a toast to his good fortune, look back with some satisfaction on what has been accomplished, and thank those shipmates who made the long voyage possible.

*Battleship Sailor* could not, of course, have been written without them. In reconstructing the battleship navy of forty years ago and peopling it with real ships and real sailors, the fog of the decades was, in patches, rather dense. Lacking the extended conversations and correspondence with old shipmates and others who are knowledgeable about the navy of the 1940–41 era, I couldn't have hoped to pierce it. My heartfelt thanks go to all those listed below and to those others who provided illumination, however fitful.

A few areas of reduced visibility remain. This is inevitable in a book that deals with countless thousands of elusive facts and murky events, many not subject to verification or open at this remove to more than one interpretation. Even if my research had been as pure as the old Ivory Soap was advertised to be, there still would remain a residue of errors sufficient to give glee to critics.

Some readers may not agree with certain of my explications, criticisms, and conclusions. This book was written from the viewpoint of an enlisted man, an involved participant, and not a dispassionate observer. While I made every effort to be accurate, the opinions expressed are my own, unless otherwise noted, and must not be imputed to any former officer or enlisted man who has generously given his aid.

It was my singular good fortune that the manuscript for *Battleship Sailor* received an incisive and painstaking critique by a man I greatly admire, both as a naval officer and as a novelist: Captain Edward L. Beach, USN (Retired). Many of his suggestions were incorporated in my final draft.

In addition to Captain Beach's review, I felt it imperative that my manuscript receive the close attention of a career enlisted man with a wide-ranging knowledge of the prewar navy. That was supplied by a fellow writer and good friend of long standing, Earl E. Smith, CWO, USN (Retired). Our friendship did not stay him from extensive and vigorous criticism of the manuscript: the finished work is much the better for it.

Paul Stillwell, editor of *Air Raid: Pearl Harbor*, is another naval authority to whom I am greatly indebted. It was he who first suggested this memoir. During the course of the writing, he was enormously helpful with morale-building phone calls and correspondence, research suggestions and materials, and critical readings of chapters in the draft stage. He has certainly proved to be a true friend.

To lend balance to this account of the enlisted service, it was important also to get the views of a career officer who had served in the *California* in 1940–41. Here I was fortunate to have the full cooperation of Captain Cary H. Hall, USN (Retired), then an ensign and Sixth (Port) Division officer. His many long, articulate letters gave me much insight into both the life of a junior officer and the organization and operation of a battleship. His critical reading of manuscript sections relating to gunnery and other technical matters was invaluable, as were his more general comments.

In recreating the radioman's world-within-a-world, I badly needed— and received—the enthusiastic help of a number of my radio gang mates. They gave detailed answers to my interminable questions, and furnished reminiscences, descriptions, drawings, war diaries, photos, and other memorabilia. This is a tribute to the contributions of John R. Mazeau, David Rosenberg, Julian Rossnick, Joel E. Bachner, and Vernon L. Leuttinger.

Thomas N. Hall, a former radioman who was at Kaneohe Bay on 7 December 1941, provided additional background on fleet communications and naval aviation.

Many other *California* crewmen of various divisions and ratings shared their memories, anecdotes, photos, and assorted materials. Foremost among them were John H. McGoran; William A. Lynch, CWO (MC), USN (Retired); David C. Moser, CY, USN (Retired); Clair E. Boggs, CWO, USN (Retired); Gordon H. McCrea; Robert H. ("Rebel") Boulton; Richard A. Donahue; Richard G. ("Smitty") Smith; John D. Grabansky, CSM, USN (Retired); Howard H. Juhl; and Captain Raymond C. Hohenstein, ChC, USN (Retired).

I was able to contact the above shipmates (and a number of others) through two organizations of which I am proud to be a member: the Pearl Harbor Survivors Association and the USS *California* Reunion Association. PHSA provided me with a current roster of the nearly 200 members who were serving in Battleship Forty-four at the time of the attack on Pearl Harbor; PHSA's official publication, the "Pearl Harbor GRAM," was an additional source of contacts and background information. Harold Bean, national chairman of the reunion association, sent along a membership roster. My thanks also to Captain Roy C. Smith, III, USNR (Retired), director of publications of the U.S. Naval Academy Alumni Association, who forwarded letters to officers I had served with and provided official biographies and data sheets.

I have read few recent navy memoirs or histories where tribute is not paid to Dr. Dean C. Allard, head of the Operational Archives Branch of the Naval Historical Center in Washington, D.C. My book is no exception. Dr. Allard promptly and graciously supplied ship's histories and other battleship materials, along with copies of the *California*'s action and survivors' reports.

Richard T. Speer and John C. Reilly of the Ships' Histories Branch of the Naval Historical Center provided the operational history of the *California* for her thirty-eight-year career, among other background data. Especially welcome was Mr. Speer's correction of an old myth about my battleship's change of name and hull number.

Deck logs for the entire period I was on board were supplied through the good offices of Dr. Timothy K. Nenninger of the Navy and Old Army Branch of the National Archives. I was also given access to logs of the *Neosho* and the *Louisville*. Thanks to all of these logs, I was able to construct an exact chronology of my travels and to verify many names and events.

Dr. Thomas C. Hone loaned me his precious copy of the *California*'s War Damage Report of 28 November 1942—essential to my account of the Pearl Harbor attack and its aftermath—and was generous with research suggestions and encouragement. A fellow writer, James Chinello, trusted me with a folder of research he had compiled on the "Prune Barge." My thanks, too, to novelist Darryl Ponicsan for his comments and criticism.

My tribute to the ferry *Kalakala* was made possible by Captain Robert E. Matson (Retired) of the Puget Sound Maritime Historical Society, a former mate and master of the "Flying Bird," who shared his comprehensive knowledge and his files.

At Puget Sound Naval Shipyard, Lieutenant Commander D. P. Mahoney, USN, the public affairs officer, gave me a tour of the navy yard and made copies of battleship photos from his archives.

Just outside the main gate, at the Bremerton Armed Forces YMCA, Mr. Robert C. Wick, the executive director, helped me find some remarkable primary-source materials and period photographs.

Lieutenant Commander John A. Marchi, public information officer at the San Diego Naval Training Center, brought me boxfuls of historical materials, loaned me old photographs, and reintroduced me to the station's excellent library.

Other period photos and background data were searched out and loaned by the Port of Long Beach and the San Pedro Maritime Museum.

My last acknowledgment, and surely the most important one, is reserved for my lovely and charming French-American wife, Rita Jeanette (née Bolduc). During the alternate frenzies and woes of research and writing, she smiled bravely and never faltered in her patience, inspiration, and (most admirable of qualities in a writer's mate) total partisanship. She is the friend of my heart. It is my good fortune to have shared these later years with *ma belle femme sans pareil.*

# Battleship Sailor

# CHAPTER 1

# Liberty of 5 December 1941

*O'ahu maka 'ewa 'ewa*
*("Unfriendly are the eyes of the people of Oahu")*

"Let's go ashore, Ted!"

My best friend, Radioman Third Class M. G. (Johnny) Johnson, was beckoning impatiently from the watertight door to the after-radiomen's compartment on the third deck of the battleship *California*.

I hesitated. The sparsely furnished compartment, in its many coats of white paint, was about as inviting as a hospital ward. Being below the waterline, it lacked ports and an adequate supply of fresh air. The temperature was a humid ninety degrees, and I was sweating.

"Damn, Johnny, I'm broke." We had been in Pearl Harbor with four of the other battleships since 28 November, and several liberties had reduced me to near insolvency. I knew Johnson was in the same condition; our wristwatches and rings were already locked up in a Honolulu pawnshop.

"Don't worry, buddy. We'll score somewhere. Let's go!"

One of the great "operators" of the radio gang, he knew at least two girls on the beach who would loan him small amounts of money—an unparalleled achievement for a navy enlisted man in the Honolulu of 1941. Since I was facing an afternoon watch on Saturday and the dreaded 0100-to-breakfast midwatch on Sunday morning, I quickly agreed.

Urged on to greater speed by Johnson, I put on my best white jumper and trousers and a pair of spit-shined black dress shoes. To get off a battleship, especially the flagship of Vice Admiral William S. Pye's Battle Force, one had to be sharp and immaculate in person and uniform. This procedure required hours of effort, so I always kept a complete uniform pressed, shined, and ready for liberty (or captain's) inspection. Making sure that my neckerchief was tied with ends even, that its square knot was aligned like a gunsight with the V of the jumper, and that my white hat was perfectly round and square over the eyebrows, boot-camp fashion, I splashed on some Old Spice, shoved a package of Lucky Strikes inside one sock, and folded my wallet over the waistband of my pants (undress whites had no pockets). Double-timing along the third-deck passageway and up two starboard ladders, I passed the ship's gedunk stand and messing compartments and caught up with Johnson as he stepped onto the quarter-deck.

Mustering with the early liberty party, a privilege accorded members of the radio gang and other divisions that had to stand 24-hour-a-day watches, we stood at rigid attention for the inspection of the junior officer of the watch. More often in the late months of 1941, with wars raging on several continents and the armed forces belatedly expanding, the JOOW was not a sword-straight Annapolis graduate but a reserve officer, known disparagingly in the privacy of the crew's quarters as a ninety-day wonder. If we had seen how these officers were treated in the wardroom, we might have felt a certain amount of sympathy for them. But centuries of British and American naval tradition had placed a gulf between officers and men in the battleship navy, and there was little or no personal contact. The USNR junior officers had just as much authority as the career navy types to put a man on report or turn him back from liberty for the slightest deviation from a shipshape appearance, and they were, accordingly, just as feared. Perhaps more so, since their lack of familiarity with navy regulations and the unwritten customs of the navy made them somewhat unpredictable.

Armed with liberty cards and twelve hours of freedom (liberty expired at 0100 for all enlisted men except chiefs and first-class petty officers), Johnson and I snapped stiff-armed salutes at the JOOW.

"Request permission to leave the ship, sir."

"Permission granted."

Facing smartly toward the stern we repeated the salute to the national ensign at its flagstaff, descended the accommodation ladder single file, and stepped into the waiting motor launch. As we shoved off, a buzz of eager, profane conversation replaced the stilted formality of the quarterdeck. At a safe distance, Johnson and I shifted the knots of our neckerchiefs toward our throats and crushed and bent our white hats before placing them on the back of our heads for a properly salty appearance. Our only concern now was to avoid the stray wave that slopped aboard the fifty-foot open launch when it intercepted the wake of a passing boat and fell heavily into a trough.

As we headed across the channel from Ford Island and the seven moored battleships, I must have looked back. I always did. On board the *California*, all was unforgiving white steel, crowded hot quarters, and continuous cold regimentation. Watches, working parties, cleaning details, drills, musters, inspections, battle stations—all were accompanied by boatswain's whistles, bugle calls, and the peremptory voice of the ship's public-address system calling out, "Now hear this. . . ." On a ship the length of two football fields with end-zone yardage to spare, it was difficult for a radioman and nearly impossible for a seaman to find a corner where he could relax, read, write letters, or simply be alone. Always, one was pursued from space to space by the impersonal directives of the executive officer's plan of the day and the impersonal commands of the authority figures near or remote, in chevrons or braid, who were delegated to enforce the plan.

But viewed from a liberty boat the *California*, with her raked clipper bow, appeared a thing of strange masculine beauty. The piled-up superstructure of command was almost perfectly balanced by the two tall cage masts that housed the main- and secondary-battery directors and by the long-barreled fourteen-inch guns in their superimposed, fore-and-aft turrets. Indeed, it was the guns that put it all in perspective. No one was on board the *California* to see the world or learn a trade or "punch his ticket" for the next higher rank. From the lowliest Negro mess attendant to the captain himself, everyone was in the service of the guns. All the machinery and equipment the men operated, all the braid and chevrons and regimentation, were dedicated to one task: loading, elevating, and training the twelve gargantuan rifle barrels, so that they would speak with tongues of orange flame and shattering sound, hurling 1,500-pound projectiles across the horizon in lofty, decaying curves to descend upon, penetrate, and destroy the ships of the enemy. To speak death was the purpose of the *California* and her sisters of the battle line, and there was an austere, terrible beauty about it all that both fascinated and repelled me.* Perhaps young men of my background have always secretly loved the violence they were taught to hate.

Moored to quays on the other side of Ford Island, where the carriers usually tied up, were a couple of antiquated four-stacker light cruisers (easily recognizable by their bulky top foremast structures and incongruously short forecastle decks), a seaplane tender, and the pre–World War I battleship *Utah*, now reduced to the ignominy of being a "mobile target ship." Beyond them in East and Middle Lochs, two of the several spreading branches of tree-shaped Pearl Harbor, dozens of destroyers were nested in groups of three to five, alone or alongside their tenders. Near Aiea was a new, square-sterned light cruiser; off the Ford Island runway, a large red cross identified the white hospital ship *Solace*.

The lochs, in fact, were awash with ships. It was almost as if the admirals were running a clearance sale on navy ships and had positioned them for the quick inspection of potential buyers. (I did not know it then, of course,

---

*The *California* had a strange nickname for a fighting ship whose purpose it was to speak death. California was primarily an agricultural state when its namesake was commissioned in 1921, and one of its major crops was plums, especially the varieties that become prunes when dried. There being many native Californians in the commissioning drafts from Yerba Buena Naval Training Station, the ship's company became known in the fleet as prune pickers. By extension, The *California* became the "Prune Barge." I am indebted to David C. Moser, who served in the *California* from 1934 to 1938, for this explanation.

but an Imperial Japanese Navy spy named Takeo Yoshikawa, who posed as playboy Vice-Consul Tadashi Morimura at the Japanese consulate on Nuuanu Avenue in Honolulu, was observing the lochs from Aiea Heights.) As a sixteen-month veteran, I had been in the navy long enough to admire the skill with which more than ninety ships had been wedged into such close quarters. I had not been in the navy long enough, nor had I read enough naval history, to ask the obvious questions: Why were they innocent of any type of underwater protection? What kind of miracle would be required to extricate them in an emergency? In any event, we enlisted men left such abstruse matters to our officers. To us a crowded harbor meant one thing: Honolulu would be jammed with sailors on liberty.

Our boat was soon opposite the long expanse of Ten-Ten Dock, where the minelayer *Oglala*, a veteran of the North Sea mine barrage of World War I, was tied up outboard of another boxy-sterned light cruiser, the *Helena*. Dead ahead of Ten-Ten, in Dry Dock Number One, could be seen the graceful tripod masts of the battleship *Pennsylvania*, which now served in name only as the flagship of the Pacific Fleet. Admirals, I had learned, took pride in planting their flags on seagoing ships, whether they actually went to sea in them or not. Next up were the various-sized docks of the repair basin, where the yardbirds (our term for the usually disheveled and customarily disliked shipyard workers) were assaulting a couple of "tin cans" and other assorted craft.

Passing the marine railway, we entered narrow, congested Southeast Loch, more a misshapen shoot than a real branch of the harbor. Above us, the elongated mass of the giant hammerhead crane, a Pearl Harbor landmark painted in black-and-white checks, dwarfed the masts of the cruisers and destroyers tied up at the two finger piers below. Off the port bow now, the tall cylinder of the diving tower, where submarine sailors took their escape training, marked the location of the sub base. Set among coconut palms and tropical shrubbery, the handsome, U-shaped administration building was the headquarters of Admiral Husband E. Kimmel, commander in chief of the Pacific Fleet. Kimmel I remembered only as a tall, rather red-faced man of Prussian bearing, who had swept by one hot Saturday morning on the quarterdeck while I stood at attention, praying he would pay me no heed.

One thing I had learned about admirals was that they were unpredictable. Often an admiral would stop the stern parade of four-, three-, two-and-a-half-, two-, and one-stripers (which, being a country boy, I couldn't help likening to a mother quail leading her brood to the water hole), select some hapless young sailor, and do a little morale building.

"Well, son," the admiral would say in a bluff, quasi-friendly tone, "how do you like the navy?"

This aerial photo of Southeast Loch, looking toward the submarine base, was taken in October 1941. Behind the U-shaped administration building are the Pearl Harbor tank farm and Makalapa Crater; across the channel to the right is the Merry Point Landing. (Photo courtesy of the National Archives.)

Suddenly the center of all attention, keenly aware of the admiral's lofty rank, the captain's piercing gaze, and the chief yeoman's pencil poised over his notebook, the chosen one would flush, clear his throat, and stammer, "I like the navy real well—sir!" My secret dream was that one day an enlisted man far more foolhardy than I would fearlessly reply: "Since you asked me, sir, I'll tell you. I don't like the navy worth a shit. Sir!" But, of course, no one ever did.

As we came opposite three gaunt black submarines in their slips, our coxswain reduced speed for the approach to Merry Point Landing. Dozens of other launches and whaleboats were bringing liberty parties to the landing, and we had to wait our turn. No sooner had the young seaman second hit the dock with the bow painter than we were pouring over the gunwale and rushing past the nearby officer's club toward the main gate.

Now the only remaining hurdle was the officious marine guard unit. Although the "gyrenes" were much more interested in who was entering the yard than in the hundreds who were leaving, we straightened our white hats as a routine precaution. We considered the marines, at least those assigned to ships and navy bases, as little more than nattily outfitted cops, and our feeling about the police was the universal one of youth. Even in the *California*, we had little or no contact with the Seventh Division of Marines. While they looked resplendent at inspections in their "seagoing bellhop" dress uniforms, they were still cops. When they weren't guarding the brig or doing rifle drill, they seemed to spend most of their time standing around practicing the marine look: erect, hard eyed, thin lipped, hostile. Or so we thought.

But on this day we passed through the gate without a glance from the guards, and now Honolulu was only eight or nine miles away along a narrow, two-lane highway. The thrifty and the indigent among us took one of the small city buses drawn up near the gate. But Johnson never considered himself indigent, except temporarily, and we followed our usual practice of taking one of the numerous waiting cabs. These were large, American-built sedans, sometimes equipped with jump seats, that took five or seven in style at dangerous speed to downtown Honolulu for twenty-five cents a passenger. The cab drivers were usually entrepreneurs who took great pride in their automobiles. While their taxis were not often new, they were certainly the most expensive and showy secondhand cars they could afford, nothing like the rattletrap Model A Fords that Hollywood filmmakers pass off as Honolulu cabs.

The first few miles of the highway were edged by fields of tall, pale-green sugarcane on the inland side. To the north and east were the sharp ridges and plunging escarpments of the purple Koolau Range, mantled in ever-present cumulus clouds blown by the northeast trade winds. On the ocean side were the Hickam Field gate and, a little farther along, a rutted road that led to John Rodgers Airport, where I had been taking flying lessons whenever I could afford them. A long stretch of low, dreary flatlands gave way to the marshes of the Keehi Lagoon. The tracks of the narrow-gage Oahu Railway (long ago dubbed the Toonerville Trolley) paralleled the highway into Honolulu.

After a few near misses as our cabbie swerved out to pass slower cars and avoided onrushing traffic with scant feet to spare, we entered the dreary suburbs of west Honolulu, which were crowded with lean-tos with tin roofs, scroungy one- and two-story Japanese business establishments, small apartment houses without a single ninety-degree angle, and the towering Dole Pineapple Cannery. The latter was not looked upon with any great favor, for one of the featured soft drinks at the ship's gedunk stand was Dole Pineapple Juice at an exorbitant ten cents per can.

The green triangle of Aala Park and the low, arched bridge over the Nuuanu Stream (known to us simply as the canal) announced that we were on the edge of the Hotel Street honkytonk district, the focus of a sailor's few free hours. We were there not by choice but by necessity—on my first few liberties in Honolulu fourteen months before, I had learned that enlisted navy men ranked at the nadir of the island's social scale. Growing up poor in a small town in northern California's gold-rush country, I had certainly been made aware of social classes. But I had never faced anything remotely like Honolulu's blank, impenetrable discrimination. The poverty of my foster-parents, while occasionally embarrassing and even humiliating to a sensitive teenager, had been no real impediment to making the high-school honor society, earning a varsity letter in sports, winning election to several student offices, being invited to join the DeMolay Order, and dating several girls of families better off than mine. On liberty in West Coast ports, I had felt free to go any place I chose and had, in the process, met several attractive young ladies. But here in this American territory I was, in effect, restricted to social contact with women in a few clip-joint areas, simply because I was wearing the white uniform of a navy enlisted man. I had heard, and believed, that a few East Coast navy towns like Norfolk, Virginia, still had signs posted saying Dogs and Sailors Keep Off the Grass. Here on Oahu, no signs were necessary to keep us in our place. When I had griped about it to Johnson many months before, he laughed and offered some comforting words.

"Don't take it personally, Ted. It's the goddam system. This has always been a lousy town for sailors. They've been coming here for a hundred years—and that's how long the big fat mammas have been warning their daughters to stay away from them.

"Not to worry, kid. One thing you'll learn if you stick around with me is that a real operator can score anywhere. All he needs is time." He shrugged expressively. "Here there's no time. Hell, we have to be back at the landing before the action really begins. So we make the best of what's available."

Accordingly, we appraised the scene and gaged the possibilities as we made our stop-and-go way down Hotel Street. On this particular Friday afternoon, they did not look promising. Hordes of hats in navy white and army drab were already bobbing up and down the street. Three more battleships had just arrived at Pearl that morning from training maneuvers, along with their destroyer and cruiser screen. In port now were all the ships of the battle line with the exception of the *Colorado*, known throughout the fleet as a hard-luck ship, but currently enjoying a stateside overhaul at Puget Sound Navy Yard. Two of the fleet's three carriers, the *Enterprise* and *Lexington*, were somewhere at sea, but this was customary when the battleships were in port. (The *Saratoga* was in San Diego for repairs.)

We had little contact with carrier sailors; and most of us had never even heard of Vice Admiral William F. Halsey, Commander Aircraft Battle Force. To a real battleship man, the bulky carriers were clumsy floating airfields of a surpassing ugliness dictated by their function, not to be remotely compared to the graceful lines of a real fighting ship with a fourteen- or sixteen-inch main battery. The sailors on these outlandish "birdfarms" were the "brown-shoe navy" and were considered little better than landlubbers.

Our cab ride ended at the Army-Navy YMCA on the upper reaches of Hotel Street, near the unlovely bulk of the rococo Iolani Palace and the Italianate columns and campanile of the Judiciary Building. Fronting the latter was an island landmark known to all sailors: the heroically sized statue of Kamehameha the Great, an unlettered general and statesman who had conquered and united the Hawaiian Islands 130 years before. The right arm of this noble barbarian was extended perpetually in a palm-up welcome, while the left one firmly clutched a long, barbed spear. Johnson chose to interpret such symbolism in navy fashion.

"Just like this goddamn town," he commented sourly. "One hand held out for your money, and the other ready to stab you in the back!"

The "Y" was set well back from the street in a handsome courtyard filled with palms and spreading monkeypod trees. With its second-floor archways and red-tile roof and lobby, it showed a vaguely Mediterranean influence. As always, it was filled with a milling throng of bored and unhappy servicemen. Some were standing in line at the pay phones, clustered around the front desk trying vainly to get a room, or elbow to elbow in the coffee shop having a civilian hamburger and glass of milk (the one never served aboard ship and the other rarely). Others were writing letters, reading magazines, or listlessly playing pool, Ping-Pong, or the pinball machines. Such sterile amusements held no charm for us, so

Sailors gather at the statue of King Kamehameha the Great in a 1941 photo. (Photo courtesy of the Hawaiian Tourist Bureau.)

Johnson and I headed directly across the street for our first drink at the Black Cat Cafe.

Four decades later, mention of that cafe still brings a gleam of recognition and perhaps an oath from Old Navy hands. Long before James Jones and other lesser writers made its name a byword for the Hotel Street of 1940–41, the Black Cat was notorious throughout the Pacific Fleet. Its single claim to fame was its fortuitous location near the YMCA, where most of the sailors and soldiers began liberty. For that first drink—and, late in the evening, the last one— it was simply the most convenient place to go.

Housed in a rickety wooden building with a plastered false front, the Black Cat was open to the street, in the tropical style. Two heavy black metal grates closed it off at night. Inside, the drinks were watered, the bartenders uniformly insolent, the decor cheap and barren, the ambiance malignant. (I still recall with delight an evening late in 1941 when a group of *California* sailors revolted and dismantled the Black Cat—or as much as could be managed before the arrival of the shore patrol forced a strategic retreat.)

That afternoon of 5 December, Johnson and I were alone, only because most of our fellow radiomen were broke, too, after several consecutive liberties. Usually, half a dozen or more would join us for the first beer or highball and at least the first reconnaissance of Hotel Street or Waikiki; for Johnson was the type of male (a chauvinistic one by the standards of the 1980s) who was both successful with women and respected by other men. Born in Seattle in 1914 of Swedish-American parents, he was a rangy six feet two inches and weighed around 190 pounds. He was handsome but not "pretty"; masculine but not *macho*; outgoing and ebullient; occasionally depressed but never withdrawn. About him on many memorable liberties in West Coast ports was always the aura of great expectations, adventures to come, promises to keep. He attracted as leader a coterie of younger, less sophisticated sailors who admired his sangfroid, his reckless, live-for-today philosophy, his exploits with women; and, because they wanted to be like him, they aped his rapid, slurred speech patterns and animated gestures and mannerisms.

On this liberty, Johnson was more subdued than usual. He had been so, in fact, since the *California* had left Long Beach for Pearl Harbor on 1 November. We had had a glorious three weeks in San Francisco and southern California—our reward for six months of rugged duty in the Hawaiian Islands. Johnny's regular girl friend, Eleanor, had come down from Seattle to share an apartment with him, first on Pine Street in "the City," and then on Cherry Avenue in Long Beach. (Since she had brought two friends along who were anything but virginal, the latter address occasioned a good deal of ribald humor.) Six or eight members of the radio gang, myself included, had made these cheaply furnished dwellings our

Author, left, with M. G. (Johnny) Johnson at Waikiki in the last photograph taken before the Japanese attack on Pearl Harbor.

home away from the ship. Eleanor was a gracious hostess who treated us all like kid brothers, and the parties went on day and night. They were fueled by periodically taking up a collection and dispatching a "working party" to a nearby liquor store or market. Those who weren't old enough to buy liquor in California were expected to do their part by "policing up the area". At 0600 every morning Johnson himself sounded reveille for those celebrants who had to return to the ship. Rousing ourselves groggily from sofas, chairs or, if late arrivals, the floor, we would splash cold water on our faces, struggle into our dress-blue jumpers, and share a cup of "joe" with Johnson before hitting the street.

When the *California* got under way with the heavy cruiser *Salt Lake City* (CA 25) for Pearl Harbor that November morning, most of us assumed we would be back soon. But Johnson, inexplicably, knew better. Although liberty expired at midnight, he had received an extension on some such grounds as illness in the family and arrived with the shore-patrol boat at 0830, shortly before we weighed anchor.

The party on Cherry Avenue had gone on all night and my friend was not sober. We stood on the quarterdeck, breathing in the unforgettably pungent smell of confined saltwater and crude oil and watching the boxy, cream skyline of Long Beach slowly recede to the north. We were leaving "the land of the sundown sea" for harsher duty, and I savored the memories while regretting the need. My friend's feelings were much stronger. As we approached the low gray breakwater, I could see a fog bank ahead, thick and menacing. But Johnson kept staring back at the silhouette of the city, wispy and almost ethereal in the morning overcast, as if memorizing it for all time. Suddenly he spoke in a low emotional voice.

"God damn it, Ted," he said. "There it goes. We're leaving, God damn it, and I'm never gonna see the States again."

The utter finality of his words and the intensity with which they were spoken left me speechless for a long moment. I attributed them to strong drink and the sadness of parting from a woman who loved him deeply. "Aw, Johnny," I said awkwardly, groping for expression. "Of course we'll see the States again. Hell, we'll be back in a few months—maybe sooner."

He took one last look, shook his head, and turned toward the door that led to the main deck. "No," he said. "Not me. I won't."

I thought about his remarkable precognition many times afterward, but not, I must admit, on 5 December 1941. Finishing a quick highball at the Black Cat, we plunged into the white and khaki tide along Hotel. Since the narrow sidewalks accommodated but three abreast, progress was only possible by squeezing sideways in single file or by stepping out into the street.

The street was only a little wider than a respectable alley in most cities in the States, comprising one cramped lane of traffic and one of parking on either side. Leaning against each other for support along its dreary meandering route were ancient two- and three-story buildings. Relief of a sort was provided by an occasional Oriental motif, as well as the variety of color schemes: off-whites, scuffed browns and beiges, bilge greens and tenement-kitchen yellows, with battleship gray predominating. Scuttlebutt had it that gray paint was favored because it was provided by the yardbirds in exchange for beer and other necessities.

Opening off the uneven sidewalks were narrow-fronted business establishments offering every type of profit-making escapism an enterprising, multi-cultural citizenry could devise for its protectors in the apparent absence of inhibitions or civic restraints. Side by side were dim bars with jukeboxes blaring the latest stateside and *hapahaole* hits (Hawaiian music modified to American tastes); photo studios where one could pose with a dusky island lass in a grass skirt before a painted backdrop of Waikiki and Diamond Head; trinket and souvenir counters; shooting galleries operated by White Russian refugees from Shanghai; jewelry and clothing stores; barber shops where flirtatious Japanese girls wielded the clippers; tattoo parlors whose tiny windows were filled with sketches of designs strange and wondrous, from a simple "Death Before Dishonor" to a dagger-pierced heart to a clipper ship under full sail; peep-show and pinball-machine joints; *saimin* ("pork and noodles") stands; even an occasional respectable restaurant like Wo Fats, with its green-tiled pagoda tower. And pervading all this was the unique smell of Honolulu: a malodorous infusion of decaying pork, overripe fish, and a variety of pungent spices unknown to Western nostrils.

On the second and third floors, reached by narrow wooden stairways, were the offices of small, secretive businesses and professions, massage parlors advertising their electrical delights in gaudy neon, and hotels of two types. One kind actually rented rooms by the day or week. The other had the names such establishments always have: Rex, Ritz, Anchor, New Senator. If still in doubt about what these hotels had to offer, one had only to follow the uniforms.

The Hotel Street of late 1941 cannot be compared to the Hotel Street of the Korean and Vietnam wars. It cannot even be compared to the Hotel Street of 1942, when martial law snuffed out such artificial and venal gaiety as it possessed. The rickety old buildings are mostly still standing, looking even more forlorn; and the mean and narrow streets down which we had to go are mean and narrow still. But the mood of those last days of peace can never be recaptured. There was a boisterous innocence about it all in the Honolulu of 1941, a brash and blind naiveté which assumed without question that the necessary steps were being taken to ensure that all would be well. Children of the Great Depression, most of us yet trusted and believed in our leaders, from our jaunty and aristocratic president, elevated by the working classes to near godlike stature, to the beribboned officers who commanded us, inheritors of a British caste system and an American naval tradition of valor and victory. We had been taught not only that the United States had never lost a war, but also that the country had committed few sins in winning a continent (frozen Canada and corrupt Mexico sharing it at our sufferance). And most of us believed. Despite the poverty, misery, and injustice of the 1930s, we thought we were citizens of the best country in the world. Our chauvinism, although incoherent and nonverbal, was deep and belligerent.

To say that we underrated the Japanese as potential adversaries would be an understatement. We saw the black headlines in the Honolulu *Advertiser* and *Star-Bulletin* when we went ashore, but most of us dismissed them with a glance. Our concept of the typical Japanese male was that of an editorial cartoon from a Hearst newspaper: short, nearsighted, bowlegged, and bucktoothed, with a thirty-five-millimeter camera hanging from his neck, smiling falsely and bowing obsequiously. It was a "known fact" that all Japanese, because of their inadequate diet of rice and fishheads, had to wear thick corrective lenses that unfitted them for flying combat aircraft. It was also thought that they were a nation of slavish imitators, able only to produce shoddy replicas of American products, and that they built top-heavy ships that often capsized when launched. The idea that this pitiful island race would dare to challenge the mightiest nation in the world was ludicrous.

Hadn't a recent *March of Time* film assured us that "man for man, ship for ship and plane for plane," the U.S. Navy was the finest in the world? And if any slight doubts remained, they would have been dismissed after reading articles like "Impregnable Pearl Harbor" in a June issue of *Colliers*, written by the famous correspondent Walter Davenport. The bulk of the article was devoted to lauding the "impregnable" defenses of Pearl Harbor and Oahu, which we accepted as probable truth because he had been briefed by three of our top admirals.

It was not discussed often, but the consensus among the senior enlisted men, whom I overheard often in the petty officers' head, was that it would take about three weeks to sink the entire Japanese Navy and steam on into Tokyo Bay with all flags flying. I had no reason to doubt this opinion, believing that it had come down from the wardroom.

Just a few days before this liberty, while I was scraping at my still sparse beard with a safety razor, a stocky second-class machinist's mate announced in a loud, cheerful voice, "I predict we're going to be at war with the Japs within two weeks." Several petty officers dissented, dismissing with contempt the abilities of the "slant-eyed bastards" and their "pagoda navy," and I was quick to agree with them. Which was, perhaps, a kind of protective reaction. I was powerless to influence either the future or my role in it, and dwelling on the one or the other would have been an exercise in futility.

Johnson and I seldom discussed such subjects on our liberties. The only Oriental on my friend's mind that Friday afternoon was a certain chubby bar girl of mixed Japanese and Hawaiian descent who usually had money and was willing to make him small loans.

"Oh, Johnny, oh, Johnny, oh!" she would shriek, echoing a popular song of the thirties, "You're so good-rooking!" Johnny would graciously accept this praise, along with a five- or ten-dollar bill, and generously buy her a drink or two before finding an excuse to leave.

As we worked our way down Hotel, the hours passed in a typical way. We stopped for a drink at every likely looking bar: The Mint, Two Jacks,

Hoffman's, The New Emma, all as standardized and interchangeable as the brands of beer and liquor they dispensed.

The velvety Hawaiian night was falling fast by the time we reached River Street, the working address of many of Honolulu's ladies of the night. Curving along one bank of the Nuuanu Stream from Hotel and north again on Beretania were the crumbling business blocks whose upper floors were houses of prostitution sanctioned (unofficially, at least) for the exclusive use of the armed forces. In one of them, Johnny knew a prostitute who had a crush on him, and since we hadn't found the bar girl, he decided to pay her a visit.

We went up a flight of creaking stairs, were passed through a speakeasy-style door guarded by a 300-pound Hawaiian in a flowered shirt big enough to tent a squad of Schofield Barracks "dogfaces," and entered the reception room. The reception area—and indeed the whole establishment—was memorable for not being memorable. It was a slice of authentic Americana—as much at home in Portland, Oregon, as in Portland, Maine—transplanted intact to Honolulu. And deliberately so. For what the faceless proprietors and the madams had provided were lower-middle-class American girls in a lower-middle-class setting of chrome and red Leatherette, small metal coffee tables with simulated wood-grain finish, tall floor lamps, and chrome smoking stands—the trappings of 10,000 living rooms and cocktail lounges in the States.

The girls ranged in age from seventeen or eighteen to a rather shopworn thirty. There were blondes, brunettes, redheads—in shapes, sizes, and personalities to suit any fantasy—and they were all white, like their customers. Those who wanted to sample more "exotic fare" had to go to one of the nearby civilian houses, where the girls were mostly Oriental, with a few Hawaiians, Portuguese, and Negroes, and where, according to reports from sick bay, venereal diseases were rampant. (The girls at the army-navy houses were inspected by military doctors every week.)

Contrary to the myths that have been propagated by certain writers and Hollywood producers, there was no liquor available in these places in late 1941. The rapid buildup of the army and navy had made it necessary to eliminate that feature of gracious living many months before. The only entertainment, aside from the basic one, was provided by a large, coin-operated jukebox faced with multicolored plastic.

The atmosphere, in fact, was brisk and businesslike. A serviceman had scarcely to sit down on sofa or easy chair before the girls were paraded. Most wore only a flimsy, revealing peignoir held in place by a single snap or button; but the costumes were varied, being limited only by the need for fast ingress and egress, like a ship's quick-opening door. Although any girl was available to any customer, some girls exercised a rough kind of choice. If the reception room was crowded, one would often gravitate toward a young sailor whose looks she approved of, or who, perhaps, reminded her of someone from her home town. Thus, the older, less attractive men, unless they did something about it, would be invited by the older and less comely prostitutes in a sort of natural-selection process.

The approach itself was most direct. A girl would introduce herself by first name, sit on a customer's lap, squirm a little or run her fingers through his hair (kissing was taboo), and ask him if he would like to go to the room. Sometimes there would be a minimum of conversation in a half-bantering and, on the part of the customer, half-embarrassed, way. The girls seemed perfectly at home and wanton in this bizarre yet curiously protected atmosphere.

Once the selection was made, a girl led her customer down the hall to a tiny room that contained a bed, a nightstand, a lamp, and a stark white washbasin affixed to one wall. White lace curtains at the window provided the single feminine touch. When the door closed, the meter started running, and at the end of the appointed time—something like ten minutes—the madam or one of the huge, coal-black maids who always resembled Hattie McDaniel would rap smartly on the door and remind the occupants, "Time's up, honey."

Before that, of course, one more part of the ancient protocol had to be observed: the exchange of money. The two dollars (three dollars if extra services were rendered) would be handed over, and the girl would usually put the bills carelessly on the night stand. The money was perfectly safe, for in addition to that giant Hawaiian bouncer, there was the shore patrol, which made frequent, unannounced visits. Few minded this except the occasional drunk. In the navy of 1941 there was no talk about invasion of privacy or civil rights, it being universally accepted that an enlisted man temporarily gave up many of his constitutional freedoms when he was sworn in. Besides, the men who visited brothels knew that the same SP who policed the Hotel and River and Beretania Street houses today might be a customer tomorrow—a knowledge that dampened antagonism toward the patrol.

The girl Johnson came to see was an attractive blonde of above-medium height, a little older than most of the prostitutes, perhaps twenty-two or twenty-three. I remember her with long tangled curls, a half-crying laugh, and a generally hysterical manner. She was always willing to loan him money—but only, of course, behind the closed door of her room. Knowing that she was making more in a day than we earned in a month, Johnson was not at all troubled by this unusual role reversal. Himself the most generous of friends, he believed that the more fortunate should share with the deserving.

The madam greeted us with cautious friendliness. Her eyes, in a narrow, strained face, were old and very knowing. I felt sure she was aware of the purpose of Johnny's visits and, naturally, could not condone such a departure from the canons of her profession. But she left after a few syrupy pleasantries, and the girl came in. She fell upon Johnny with glad cries and then told him with regret, tears glistening in her pale-blue eyes, that she was booked up for the evening and wouldn't be able to see him. Sensing immediately that he was not pleased, she begged him to return the next evening.

"I've got the duty," he explained tersely.

"Then Sunday? Please?"

Maybe Sunday, he agreed; and we left as poor as we had arrived.

Even though I had been raised within the strict, even puritanical precepts of Christian Science, I did not find it shocking that my friend borrowed money from bar girls and prostitutes. I had been in the navy long enough to know very well how enlisted men were treated ashore. Considered fair game by civilians of all sorts, they were, generally, exploited without mercy. Despite the growing numbers of reserves and "duration" enlisted men, it was still essentially a volunteer, career navy in 1941. The occupation of bluejacket had not yet become respectable (even the word has a faintly derogatory ring), and one quickly learned that he had few friends away from the ship. Most of the radio gang rather approved of Johnson's single-handed "crusade" to turn the tables; he was widely regarded as a sort of carefree Robin Hood in undress whites.

What was acceptable behavior ashore, however, was quite another story on board. Informal loans were freely made among the ship's company, but the invariable rule was that they must be repaid in full on the following payday, when we lined up with our chits and received crisp greenbacks from the disbursing officer. While some "loansharking" (the practice of loaning three dollars for five, six for ten, or ten for fifteen) inevitably occurred, it was not nearly so widespread as in the army.

Perhaps the most mortal of all sins was to steal from a shipmate. Anyone caught *flagrante delicto* was fortunate to escape with a court-martial; more likely, he would be savagely beaten first. What the enlisted men had in fact evolved (and many former navy officers may be wholly unaware of it) was a code of conduct, a rough-and-ready approximation of the honor code of the military academies: I will not lie, cheat or steal; and I will not tolerate anyone who does. Though seldom discussed and never, so far as I know, reduced to written form, the code was operative, its existence proved by the remarkably few violations. The code, however, only applied to navy personnel in general and one's shipmates in particular; it was not extended to civilians. (One broad exception to the code was the act of lying to one's officers, which was considered perfectly proper and, in fact, eminently sensible.) When the navy had to lower its standards during the World War II years and those that saw our emergence as a superpower, this admirable code was diluted.

Johnson and I were now nearly out of funds, so we had two choices: to return to the ship or make our way back up Hotel looking for the Japanese girl or a shipmate who was flush. For Johnson, in his present mood, that was no choice at all. Pooling our change and a dollar bill or two, we charted a cautious course, lingering long over each twenty-five-cent bourbon and Seven-Up.

With time to kill, our conversation turned to the States and the women we had known. Since Johnson had been my tutor for the past ten or eleven months, I had little to contribute that was not already well known to him. But I was always a good listener. And my friend had a large and varied repertoire of tales of conquest from his years in Alaska, the merchant marine, and the navy. Unlike the boastful narratives of some shipmates, and with due allowance for a certain amount of embroidery for dramatic emphasis, Johnny's exploits had a ring of authenticity. Also unlike most, his stories were never coarsely offensive and seldom demeaning of their subjects. One of the secrets of his success, I realized even then, was that he genuinely liked and respected women. Even those who hung around the waterfront dives that catered to sailors responded to some essential gentleness and chivalry in his nature and became, for that little while at least, more refined in language and manner.

My friend, from whom I was learning so much of subjects not taught in schools, was, in short, a romantic. More and more his conversation turned to the times we had had on Pine Street and Cherry Avenue and the woman who had made them so memorable, now waiting for him in Seattle. I do not recall his making any plans for a future with her—it was as if he had abandoned hope of that since our conversation on the quarterdeck.

Now, in the raucous sound and confusion of a Hotel Street bar, with the jukebox playing and replaying such current favorites as "I Don't Want To Set The World On Fire," "Green Eyes," "San Antonio Rose," "Sweet Leilani," "Moonlight Serenade," and a new Bing Crosby hit called "White Christmas," we relived those days so near in time but already so much farther away than the vast expanse of ocean that lay between.

I thought about one of Eleanor's friends, a lost little sailor's girl named Mickey, who had sometimes been kind to me when there was no one else around she preferred. Only nineteen, she was already doomed by a congenital heart condition. The other friend who had accompanied Eleanor down from Seattle was good for a chuckle or two. She was collecting allotments from three battleship sailors, two of whom were crewmen in the *California*. Fortunately for her, one was in the starboard and the other the port watch. The fun came when one of them got special liberty.

The young lady from Seattle who was collecting allotments from three battleship sailors. Although some might question her survival technique, she was a kind-hearted and generous person.

Another memory was our Long Beach landlord, a worried little man who marched up to the front door at least twice a day, intent upon evicting us until invariably pacified by Johnson's outlandish promises and laughing charm. These dialogues would be conducted with the utmost civility on both sides.

"All right, Mr. Johnson," the harassed civilian finally would say. "But will you please *please* keep the noise down this time?"

"Yes, sir, Mr. Jones," Johnny would reply, a friendly hand on the man's shoulder. "I'm sure going to do that." Turning, he would shout through the open door, "Did you swabbies hear that? Now let's tune out the static and cut the gain!" Whereupon, after a discreet interval, the party would resume. . .

By 2100 hours, we were halfway to the "Y" with only enough cash left for a couple more drinks apiece and our cab fare. We had just doubled back to check out a cafe on the *mauka* ("toward the mountains") side of Hotel, when Johnson spotted a possible benefactor.

"Hey, Ted, there's the chief."

Across the street, standing in front of Wo Fats, was the only chief who mattered to us: Chief Radioman Thomas J. Reeves. There was no mistaking that commanding presence. The left arm of his white jacket was nearly solid with gold chief's insignia and at least six gold hashmarks, for the chief had served with the Atlantic Fleet during World War I.

He was one of those men of above-medium height and large frame who become corpulent but never look fat. He had a round, smooth face under a full head of thick, iron-gray hair that gave him a leonine look despite the military haircut. His piercing, hawklike eyes could have been those of a surgeon—or a riverboat gambler.

Even then he was a near-legendary figure. A pioneer during the primitive days of the arc transmitter, he was a man who scorned the easy shore billet that could have been his for one of the toughest jobs in the enlisted navy: chief in charge of the combined ship-flag radio gang, Commander Battle Force. He ran the ninety-man C-D Division with a discipline that was firm without being oppressive, and a competence that was awesome. Reputedly, he had turned down a commission of at least two stripes, which was certainly understandable, for no junior officer in a battleship had anything like his real authority and prestige. The officers of the communication division were utterly dependent upon him, and knew it. Not a single one, to my knowledge, could copy a Fox schedule (fleetwide communication system) message, operate a key even at ham radio speeds, tune a transmitter, or make repairs to the radio gear. Nor were they expected to. The chief alone decided when and where you stood watch, what your battle station was, when you went on liberty, and when you were ready for a faster radio circuit or an advance in rating. Within his division, he was more feared and respected than the captain himself.

I certainly shared that fear and respect. The chief and I both were well aware that I was not yet a truly competent radio operator by the standards of that time. So I had no gripes about the assignments he gave me: the slower, easier circuits in main radio; the daily training sessions in emergency radio; the battle station not where the action was, in radio one, but high atop the mainmast, where I copied by hand the shakily transmitted spotting reports from the observation planes for the correction of main-battery fire. I knew that just as soon as I was ready the chief would give me more responsible watch and battle-station assignments.

This night he was smoking his ever-present thick Havana as he observed the action on Hotel. Since I had never before seen the chief away from the ship, I hesitated. I felt about him as I had felt about my football coach at El Dorado High when I was a callow, uncoordinated sophomore. Nonetheless, this was one time I could not wait for Johnson to take the initiative: he had been AOL too many times and was in permanent residence on the chief's "shit-list." Summoning my eighty-six-proof courage, I waved and ran across the street to Wo Fats.

The chief greeted me in a friendly way. Suddenly, he did not seem quite so formidable. When he asked me how things were going, I explained my problem. "Well, Mason," he said without hesitation, "let me make you a small loan." He pulled a bill from one of his pockets and handed it to me. "That enough?" he asked.

I was holding a twenty-dollar bill—a third of a month's pay for a third-class petty officer. I must have stammered in explaining that I didn't need so much, that a five would be plenty.

"That's all right, Mason," he said with a wave of the hand and a flash from the large diamond ring he wore on his little finger. "Keep it."

I thanked him effusively. The chief only smiled in an amused way, like a worldly uncle observing the antics of a gawky and immature nephew who he hopes may yet amount to something.

My last words to him were to the effect that I thought he was a great guy. Looking back, I would not change the words. It was the kind of praise that this tough, self-confident yet perhaps lonely man could not have heard often, and I like to think that it rather pleased him.

Now the surprised Johnson and I had ten dollars apiece to finish our liberty, and I would have enough left over for a flying lesson at John Rodgers on Sunday. While I had had sufficient hours in a Piper Cub to solo, I felt I needed one more session practicing takeoffs and landings with the instructor before doing that.

The hectic return to the ship presented a striking contrast to our departure less than twelve hours before. The battleship navy, so meticulous about the appearance of its men when they went ashore, took a most tolerant view of their return from the perils of the port. The unwritten rule seemed to be that if a man was sober enough to somehow navigate the accommodation ladder, salute, and manage the words "Report my return aboard, Sir!" the state of his intoxication and the condition of his uniform were of little consequence. I have seen men pass the OOD who reported aboard too drunk to stand and had to be supported by their buddies; others who had to crawl up the ladder on hands and knees, coming erect with one supreme effort of will for the salute and the mumbled protocol.

The only men who got put on report were those few who were loud and insubordinate or the even fewer who tried to smuggle liquor aboard. I recall one burly master-at-arms who took sadistic pleasure in rapping a suspected smuggler sharply on the shins or rib cage (two favorite places of concealment) with his billy club. When he was rewarded with the muffled pop of a breaking bottle, a shower of glass on the quarterdeck, and a spreading stain on the embarrassed gob's uniform, his joy as he put the offender on report would have done credit to the Fallen Angel himself.

As the Cinderella hour approached, thousands of men would overflow the rough creosoted planking of the fleet landing in a scene of milling chaos. Every time a ship's launch or whaleboat was called in from the dozens standing off with engines idling, a mob of white hats surged forward struggling to get aboard. Occasionally, some unfortunate lost his balance and plunged into the dirty brown waters, but he was quickly fished out by his buddies or one of the boat crews amid cheers, jeers, and cries of "Man overboard!"

As each boat reached what seemed capacity plus ten percent, it shoved off with a clanging of bells and was replaced by another dead astern. Often there were late arrivals who leaped or hurdled into the boat across open water and fell heavily atop their shipmates to profane protest. Out in Southeast Loch, a few merrymakers raised strident voices in off-key song, but were finally silenced by repeated shouts of "Knock it off!" or "Belay that goddam racket!" from the senior petty officers.

The atmosphere, in truth, was sullen and volatile. Many of the men were frustrated and therefore hostile, with more than adequate reason. They had nothing to show for their time and money but a bellyful of cheap whiskey or green Primo beer and, perhaps, a brief encounter with a prostitute. The men, almost without exception, detested Honolulu. And they were returning to a ship that was notorious throughout the Battle Force for its severe discipline. But the very discipline that made them reluctant to return also damped the sparks that might have ignited a free-for-all limited only by the lack of punching room. Few were hostile enough or drunk enough to risk being put on report and facing the captain at mast. Consequently, I never saw anything more than mild push-and-shove incidents, more sound than fury, that were quickly resolved by cooler heads.

As we crossed the channel toward the *California*, Johnson and I took a last look around the harbor. On our starboard hand were the dim silhouettes of the battle line: the *Oklahoma* and *West Virginia* moored outboard of the *Maryland* and *Tennessee* in a tangle of masts; the *Arizona*'s tripod battle tops showing above the superstructure of the repair ship *Vestal*; and the *Nevada* bringing up the rear. The great ships were dark except for their anchor lights and a single white splash at the quarterdecks toward which the liberty boats were homing. Overhead, drifting clouds extinguished the stars. Below, it was warm, still, and oppressively humid.

"God damn it, Ted," Johnson said. "Isn't this a hell of a place to be?"

So ended our liberty of 5 December 1941. It was the last peacetime liberty I would make until VJ Day nearly four years later. But the god of war was not as kind to Melvin G. Johnson and Thomas J. Reeves. It was the last liberty they would ever make.

CHAPTER 2

# Boot Camp at San Diego

*This generation has a rendezvous with destiny.*
Franklin Delano Roosevelt

It was late on the sultry Sunday evening of 2 August 1940 when the Yellow Cab let Floyd C. (Bill) Fisher and me out at the main gate of the San Diego Naval Training Station. Our dress-blue uniforms were wrinkled and none too clean, and neither were we after fourteen hours on a Greyhound Bus from San Francisco.

Since our orders gave us until 0800 the following morning to report for duty, we could have stayed at the YMCA, but I knew I would not be able to sleep in my concern about the ordeal to come.

"You worry too much, Mason," my high-school buddy had said. Being of a pragmatic, even phlegmatic, temperament, Fisher kept his emotions carefully hidden behind a brusque exterior. But he made only a token objection, reasoning we could use the room money for something more important, like beer.

Carrying our canvas duffel bags—we had left our civilian clothes behind, on the advice of our reserve officer—we presented our papers to the duty petty officer. He eyed us suspiciously: recruits were not supposed to show up in uniform, especially at 2300 hours. Gesturing for us to remain in the anteroom of the tile-roofed sentry house, he took our papers to a desk equipped with a crook-necked lamp and studied them in obvious puzzlement. He soon reached for a phone and dialed a number. After a short conversation, he hung up and motioned us inside. "Wait here," he said shortly. "The OOD's on his way."

Apparently, the navy was not prepared for two lowly apprentice seamen of the V-3 Communication Reserve who had just volunteered for one year's active duty and were scheduled for accelerated boot camp and radio training. Fisher and I flicked lint and cigarette ashes from our blue serge uniforms, straightened our white hats, and surveyed our scuffed black shoes with some apprehension. Suppose we are unaccountably out of uniform, I thought, reasoning from a dense ignorance of navy regulations, and start our service careers in the brig? After the tension and anxiety of the two-day bus trip from the old gold-rush town of Placerville, with an overnight stop in San Francisco, anything seemed possible as we waited in the shadowy room, the object of curious stares from the petty officer and his seaman assistants.

Not daring to light a cigarette, we stood there for what seemed a long time before a car pulled up. A stern-looking ensign strode in, posing another immediate problem. Did one salute indoors or not? Being covered, I decided a salute was in order. Fisher followed suit. Evidently it was the right thing to do, for he returned them smartly, with a long sideways glance at these two strange people, neither navy nor civilian. Seating himself at the desk, he consulted briefly with the petty officer and examined our papers. Then he reached for a clipboard on the wall and shuffled through a thick file. Nodding his head, he passed an order and leaped to his feet.

"Well," he said, rubbing his hands in satisfaction at having solved the mystery. "You men are part of the V-3 radio-school detachment. Volunteers. Very well. I'm going to send you over to D Barracks with the other V-3s. We'll get you squared away in the morning." He was gone before we had a chance to salute.

So far, things had gone rather better than I had expected. We got out cigarettes and lit up. I was glad that I had a close friend with whom to share the boot-camp trial. Fisher, too, looked relieved: his rather coarse, freckled face relaxed, and he even managed a fleeting grin when I commented on the confusion our arrival had caused.

Eventually, a pickup truck arrived and we were driven to our barracks by a bored and sleepy seaman first class. "Second deck," he said as we got out. "Take any bunks." He roared away.

We were standing in front of a handsome two-story barracks in cream-colored stucco with a red-tile roof. Like the rest of the training station, so far as I had seen, its architecture was in the Spanish Mission style. We went up a double flight of stairs and entered a long room lined on either side with two-high iron bunks. By the dim light of hooded lamps, we could see that most of the bunks were occupied. Tiptoeing up and down the aisle, we finally found two adjoining lowers. I breathed a sigh of relief. How could one get into an upper bunk in a strange barracks at midnight without waking the man below?

Quickly, we got out of our clothes and eased into the bunks in our GI skivvy shirts and baggy, boxer-style shorts. Not knowing what to do with my uniform, I folded it as neatly as I could and draped it across my ditty bag.

"God damn, I'm tired," Fisher said. It was his way of admitting that the last two days had not been easy ones. "Good night, Mason." He promptly fell asleep.

I could not. I lay there in the breezeless summer night, hearing the cacophony of sounds produced by a roomful of sleeping men: snores, snorts, moans, strangling noises, even growls. Sailors, I was soon to learn, are restless sleepers. Three feet away, a man stirred, kicked at the single white blanket that covered him, turned over with a whistling sigh. I could see that he was short and powerfully built, and that he was already losing his hair. I had thought the navy was for the young, but this fellow was in his late twenties at least. What was he doing here?

And what the *hell* was I doing here? I could be home in my beloved Sierra Nevada foothills, sleeping in a bed almost as narrow as this one but downy with the comfort of the known. Even as I sentimentalized what I had so recently left, a small cold voice whispered that it was not for me.

In Placerville I could do nothing but manual labor: picking pears, pumping gas, "tailing off" a high-speed band saw in Beach's Box Factory. Or chauffeuring Dr. Claude C. Long, a San Francisco doctor and gentleman farmer, down country roads while he practiced marksmanship with a .38 revolver on the ground squirrels and chipmunks. In any event, the girl who had made all this hard and poorly rewarded labor worthwhile had married another, and I knew I would never get over her without putting many miles between us.

I thought often of alternatives, as did Fisher, who was as limited in his opportunities as I. Although we were honor students, not a single college scholarship of any kind had been offered to our high-school graduating class. Both of us joined the V-3 Communication Reserve in mid-1939, thanks partly to aggressive recruiting by a petty officer in the local unit. When the chance came to volunteer for a year's active duty in the fleet as radio operators, we decided that was a far more interesting challenge than anything Placerville had to offer.

Now, as I twisted and turned on my mattress in D Barracks, I wondered what I had let myself in for and whether I would be able to take it. But growing up in poverty during a depression has one advantage: of necessity, one acquires a certain toughness of outlook. I swore that, by God, I would endure the experience.

I had scarcely fallen asleep when I was roused by a police petty officer striding down the aisle and beating a tattoo on the bunks with his night stick. "All right, you boots, rise and shine! Reveille!"

I scrambled from my bunk and joined the exodus to the head at one end of the barracks. The fixtures, walls, and tiled floor were all spotlessly white—and the toilet stalls had no doors. No one seemed to mind but me. Somehow, I got shaved and showered and put on clean skivvies under my wrinkled uniform. There was no time to try to find an electric iron.

Nor was there time for a close look at my barracks mates until we assembled on the sidewalk to be marched to breakfast. Suddenly, I didn't feel so bad. In the ragged files were V-3 reserves in dress blues, in undress blues, even in civilian clothes. Half this motley crew, I was gratified to note, couldn't even execute the orders shouted out by the petty officer: "Right, face . . . forward, harch!"

Breakfast in a large mess hall, echoing with the clatter of utensils and the loud conversations of many sunburned recruits in undress whites and canvas leggings, was a substantial one: corn flakes, scrambled eggs, fried potatoes, coffee cake, and a red apple. Best of all was the steaming coffee sweetened with canned milk and sugar and served in handleless mugs.

Before lunch (known in navy parlance as dinner), we were marched to the dispensary for our physical examinations. We had already passed rigorous physicals to join the reserves, and these were rather cursory, although in the future we would have to face the dreaded smallpox and typhoid shots. After going through the double cafeteria line (where I was introduced to the navy's famous bean soup), we were again marched, if one could dignify it with that word, to the South Unit.

This was the infamous detention barracks, where regular navy enlistees underwent twenty-one days of purgatory without hope of salvation by priest, mother, or congressman. The satanic overseers of this place of penitence for civil sins were hard-bitten chiefs and petty officers from the fleet, men with impeccable records who, after six years at sea, knew good shore duty when they saw it and knew exactly what they were expected to do with these scrawny, undisciplined civilians. And indeed the recruits were a bedraggled lot in their wrinkled whites, their heads shaven to the scalp, their expressions piteous and woebegone. It was a tough but effective system, and the justice, for survivors, was built in. They in their turn would come back as real men-o-war's men to initiate a new crop of "spoiled mamma's boys" into the customs and traditions of the navy. And they would show just as little mercy in their performance, gleeful at times, of their duty.

Occasionally a little too gleeful. When I was at the training station, the story was still circulating of the platoon petty officer, a boatswain's mate first class, who boasted that his rifle platoon was so well disciplined it would do nothing except on his command. To prove it, he marched the platoon in column across the parade ground toward the boat channel. As the van of the column approached the rubble of stone that marked the water's edge, he gave no order to halt or change direction. Obediently, the recruits scrambled over the rocks and plunged into San Diego Bay, still at right shoulder arms and in near-perfect order.

Even by the standards of 1940, this was an excess of zeal. The boatswain's mate was court-martialed for unnecessarily risking the lives of his charges and was "busted" to coxswain. He was still drilling recruits in August and September of 1940, a tall, rawboned, red-faced third class who exceeded even my company CPO in the ferocity of his expression.

Fisher and I being members of the pampered V-3 Communication Reserve, we were spared this three-week ordeal of constant humiliation, regimentation, and close-order drill. We were only in South Unit to draw our clothing allowance. The long room we shuffled into smelled of canvas and leather and mothballs. As we passed down an assembly line of storekeepers behind a wide counter, we pushed along an ever-growing pile of clothing: dress blues; dress whites (soon to be eliminated from the prescribed outfit); undress blues and whites; white and flat hats; double-breasted peacoat; blue denim work pants and lighter-blue short-sleeved shirts; T-shirts and drawers; a black silk neckerchief. After the clothing had been issued—by "seaman's eye" with no regard for the recruit's opinions—we picked up a mattress, a pillow, two white cotton mattress covers, a pair of white wool blankets, and a stiff canvas seabag.

Finally, there was a pause while we waited in line at the stencil machine. I now learned that every item of clothing, except for the few that were provided with labels, had to be stenciled with the owner's name in black or white paint, and that the penalty for theft of another's blue jumper or peacoat was two to four years of confinement in a navy prison with dishonorable discharge. It had never occurred to me that someone would filch another's clothing—I didn't know then that only the first issue was free and one had to purchase replacements. At thirty-six dollars a month for a seaman second class, some found "midnight requisition" an irresistible temptation.

My conscience was already bothering me. As I was passing down the line, a storekeeper had given me a quick, appraising glance, muttered, "About a thirty-four," and slapped a new dress-blue jumper on top of the pile. Since I was already wearing dress blues issued by the reserve unit in Placerville, I didn't rate a second pair. I started to point this out, but his glance had shifted to Fisher.

"They gave us extra jumpers," I told Fisher. "Who should we report it to?"

My buddy gave me a scornful look. "How long you been in the reserves, Mason?" he demanded.

"You know how long. About a year."

"They ever pay you a dime for the drills and for digging that goddam rifle range at El Dorado and driving to Stockton to qualify with the thirty-ought-six?"

I had to admit I had received no reimbursement yet.

"Well, wise up. You just got a free jumper and pants and a couple of white hats and some shoes and skivvies for your pains."

His logic was compelling. Besides, there was no one to confess to. Now the operator of the stencil machine was looking up impatiently. "Your name, Mac?"

"Theodore C. Mason."

He shook his head at the stupidity of boots. "Your initials, Mac."

"What do you think the initial for Theodore is?"

"Just the goddam initials, Mac."

We glared at each other. I ought to knock you right off that stool, I thought. But that would have been most unwise. The description of the navy prison at Portsmouth, New Hampshire, had sent cold chills up and down my spine. "*T* as in Theodore," I said. "*C* as in Charles. Mason. *M-A-S-O-N*." Wordlessly, he punched out the stencil in letters three-quarters of an inch high. "Thank you, *sir*," I said.

"You better watch your temper, Mason," Fisher advised as we fell in outside. "This ain't Sheriff Smith you're dealing with."

Fisher was right, of course. In Placerville, I knew the sheriff and the town marshal. I knew the district attorney, and I knew Superior Court Judge Thompson, for it was he who had signed my adoption papers when I was six years old. These men greeted the high spirits of local boys with a tolerant smile, knowing that the growing pains of adolescence deserved nothing more than a brief informal lecture. But here at the naval training station, where I was only a last name attached to a number, the standards of conduct were vastly more strict. Were I to acquire the reputation of a troublemaker, the repercussions would follow me throughout my time in the navy. I promised myself that I would indeed watch my temper.

After supper, we spent the evening marking our gear. Every item of clothing, from handkerchiefs to blankets, had to be stenciled in the exact location prescribed in *The Bluejackets' Manual* (popularly known as the *BJM*)—a messy job that, it seemed to me, would have challenged a Picasso. Unless the stiff paper stencil was held down with uniform pressure along the yielding fabric, the paint would run under the edges and the letters would blotch.

On the first few pieces I tried, it was impossible to say with any certainty whether my name was T. C. Mason or I. D. Webdh. When I turned to Fisher for help, he was uncooperative. "God damn it, Mason, you're in the navy now. You'd better learn how to take care of yourself."

Surreptitiously wiping paint off my hands with one of my I. D. Webdh socks, I was about to try again when a hard young voice asked if he could help. A swarthy, pug-nosed V-3 of below-medium height but muscular physique was at my elbow. Gratefully, I let him take over the stencil, the brushes, and the pots of paint, and watched him perform this difficult chore with casual ease. When I expressed my admiration, he shrugged. "I've been working with my hands ever since I was a kid in Honolulu," he said. "You name it, I've done it."

"My name's Ted. You know the rest."

"Jack Villafranca," he responded, and we shook hands.

"Maybe you can return the favor," he continued. "I don't know the goddam code, and I'm gonna need some help."

"You know something? I don't know the goddam code, either."

We looked at each other in some dismay. There was nothing to do but laugh. "Hell," I said. "We'll help each other."

"You know any theory?" he demanded.

I explained that I had taken part of a correspondence course in radio from National Schools in Los Angeles, and that if I could be of any help I would. "That's a deal," he said, and I had made a friend.

After the clothes were stenciled, our company petty officer, a salty bosun's mate first, showed us how to roll them so that they were smooth and taut and secured with cotton stops, the stenciled name on the outside.

"Roll 'em right," he counseled, "and you can sail from here to Shanghai and make liberty in a uniform that ain't got a wrinkle."

Again, the procedure was easier to observe than to emulate, and Villa-franca had to help me. By the time we had rolled my clothes and stowed them within the narrow cylinder of the seabag, in horizontal layers at right angles to one another, there was nothing I would not have done for my new buddy.

After chow the next morning we were mustered in front of the barracks in undress whites, duty belts, and leggings. A chief boatswain's mate of ordinary stature but formidable mien, his mouth in a perpetual scowl, stood before us. Hands clasped behind him, rocking back and forth on his heels, he surveyed the group of some thirty reserves with a stony eye. I got the distinct feeling he would rather have been at sea.

"Today," he said without preamble in a biting voice, "we're going to start teaching you *dih-dih-dah* people how to be military men. Sailors. Men-o-war's men." He made a coughing sound deep in his throat and I felt he had to restrain himself from spitting. "Now that looks like an impossible job from where I stand—but we're going to do our best. *Fall in on the Grinder!*"

We moved across an expanse of thick springy lawn dotted with cypresses lovingly manicured into domed cylinder shapes and went down a long flight of granite steps. At the bottom was the feared rectangle of asphalt named Preble Field after the squadron commander in the *Constitution* during the Tripolitan War, but known to generations of boots simply as the Grinder. Several companies of about one hundred men each were already hard at work executing the hoarse commands of their petty officers: "Right shoulder, *harms!*" and "Right, face, forward, *harch!*" and "Column, half left, *harch!*"

The August sun beat down through the dissipating overcast and glared off the green waters of the bordering channel. Leather heels by the thousands slapped against the pebbled tar. Sweat stains spread across white jumpers as the shouts of platoon petty officers, high and thin and peremptory, challenged the vast space of the Grinder. The government of the United States of America, in all its power and glory, in all its impersonal

The author posed for the home-town folks in front of one of the old camouflaged guns at San Diego Naval Training Station in August 1940. The guns overlook "the Grinder" in the background.

authority, was turning callow youths into disciplined men, that they might prove worthy of protecting the greatest of nations against its numerous misguided enemies. And, for the first time, one small and anxious volunteer realized what a serious business it truly was.

After such an introduction, our actual performance was anticlimactic. We fell in and formed in ranks. We dressed right. We faced to the flank, to the oblique (pronounced to rhyme with *strike*), to the rear. Finally, we were under way, more or less in step. "Column left, *harch!* . . . By the right flank, *harch!* . . . To the rear, *harch!*" The latter maneuver threw our platoon into incredible disorder, and the cream of the naval communication reserve found themselves face to face and trying to march through each other.

"As you were!" bawled the salty bosun's mate first. "Fall in!" He looked toward the chief and made an infinitely slight shrugging gesture; I could have sworn they almost smiled.

I had done everything right where many of my fellow V-3s had performed miserably in this elementary drill, and I was soon to need that slight boost to my ego. That afternoon we were marched, in undress blues and in slightly better order, to the communication building at the west end of the station, near Rosecrans Street. A large room on the second floor was equipped with an elevated desk, a blackboard and neat files of one-armed classroom-type seats on one side, and long wooden tables and companion benches on the other. Typewriters, in back-to-back rows, were bolted to each table. A set of earphones lay across each "mill." Small black-and-white photographs of a battleship, an aircraft carrier, a cruiser, a destroyer, and a submarine were the room's only concession to decor.

Behind the front desk was a radioman first class in dress blues with three red hash marks denoting at least twelve years in the navy. When we had been seated in alphabetical order, he got up, tugging at his blouse and sleeves. He was a tanned, handsome man who resembled Preston Foster, the popular he-man actor of that time.

"My name is Craighead," he said in a mellow voice with a definite Southern accent. "I'm going to be your instructor." He paused and looked us over, seemingly a little nervous. "This is a new thing for us, teaching V-3s, so we'll have to sort of feel our way for a few days. First of all, how many of you are ham operators?"

Six or seven hands shot up.

He nodded. "Good. How many of you know the code?"

Another eight or ten hands.

"All right. Now, the rest of you. Who knows any radio theory?"

My hand went up, at last, and Fisher's, and four or five others. That left about half a dozen pilgrims, my new friend Villafranca included, who lacked even that basic preparation.

In this photo, probably taken about 1930, regular navy trainees learn international Morse in a classroom similar to one the author attended (except for the unshipshape clutter in the left foreground). Even photos of five types of combat ships on the wall are identical. But the atmosphere was less regimented in the author's classroom: no "boots" were worn. (Photo courtesy of the San Diego Naval Training Center.)

Craighead whistled softly. "Well, I see I have my work cut out for me. I hope no one is going to tell me he can't use a mill—a typewriter."

Someone raised his hand (I think it was Fisher) and asked whether "hunt-and-peck" typing was okay.

"If you don't know the touch system, it's too late to learn it here," Craighead said. "I've seen a lot of good operators get by with two or three fingers. If you don't have a system, the thing is to develop one—and fast."

"One more thing," he added. "How many of you know how to handle a service rifle?"

My hand went aloft again, along with Fisher's and Villafranca's and six or eight others.

"That's something," our instructor observed. "If you can't learn the code, at least we can make deckhands out of you."

He smiled and we understood he intended the remark as a joke. We laughed appreciatively. He was the first superior we had met who was not deadly serious. That augured well for the future.

The hams and other advanced students were soon sent to the other side of the room, where their speed could be checked with the automatic code machine to which earphones were connected.

Craighead got out a hand telegraph key hooked up to an oscillator and gave it a few tentative taps.

"Before we begin," he said soberly, "I'm going to scare the hell out of you."

He gestured toward the V-3s at the typewriters. "The hotshots over there have got it made. This is only a review for them. But you people have a problem. I had four months to get up to Fox-schedule speed. You untrained reserves have only got six weeks.

"Six weeks, men." He looked from face to attentive face. "At the end of that time, you've got to copy eighteen words a minute. You've got to send fifteen—and pass a test in radio theory, too.

"Now if you flunk, you don't get your third-class crow and you don't become a watch stander on NPM Fox. You go to the fleet as seaman-second radio strikers."

It was very quiet on our side of the room. "In case you haven't guessed," our instructor continued, "a striker is the lowest form of life in a radio gang. Your duty is to make coffee. You wake up the watch standers. You run messages. And you do mess-cook duty. If you have any time left over, you just might get a shot at the code machine."

We looked at each other in dismay. One reserve threw up his hands in a gesture of surrender. Craighead smiled.

"Men, if you work like hell for the next six weeks, some of you will make third class. Some of you won't. But if you want that crow, you'd better damn well give it your best effort."

I wiped my sweaty palms off on my pants. It seemed impossible to go from zero to eighteen wpm in a month and a half. Craighead made the prospect look dubious, and he was an old pro. For a moment, I considered whether I should even try. But of course I would try. What alternative was there?

"Okay, let's turn to," the first class said, his thumb and first two fingers resting lightly on the round black button of the key. "We'll take five letters at a time. Now A is *dit-dah*—a dot and a dash—and it sounds like this. . ."

Life now fell into a routine so Spartan that even today the memory of it can evoke apprehension and head-shaking disbelief. Reveille was at 0545, followed by a hasty breakfast and cleaning detail. By 0730 we mustered and marched to the Grinder for the manual of arms, performed with real if unloaded bolt-action Springfield '03s (known in the service as the United States magazine rifle, model of 1903). Before long we advanced from the school of the squad to the school of the platoon. In the afternoons, four hours at the communication building for the agonizing struggle with international Morse, with periodic breaks for instruction in navy signal systems, procedure signs ("prosigns"), and message forms. So furious was my concentration that the concave, Bakelite ear pieces of my headset were continuously wet with perspiration, and my stomach constantly tied up in very nonregulation knots.

The early supper left several free hours in the barracks, which I could not afford to waste in loafing or skylarking. First, there was the study of "Subjects All Enlisted Men Should Know—A to N" in our Bible, *The Bluejackets' Manual.* I didn't mind the book's two-dollar price. When I reported for active duty, I had been automatically advanced to seaman second and was now making thirty-six dollars a month, nearly all clear money. But I did mind the time it took to pore over some 180 pages of densely compressed type, written in a stern, exhortatory style.

Putting aside the *BJM* after an hour or so, I would next tackle the navy courses in radio fundamentals—voltage, current and resistance, capacity and inductance, circuits, and principles of radio communication—that we were expected to master on our own. Usually, I shared my limited knowledge with Villafranca and several others over a study session at the barrack's only table. If I was stumped, which was often, we would call on one of the hams for an informal lecture.

And in every waking moment that was not otherwise occupied, I labored to fix forever in my mind the arbitrary symbols of Samuel F. B. Morse's strange, ingenious language. Shaving and showering, shining shoes, or washing clothes in a galvanized iron bucket with my scrub brush and the yellow bars of harsh GI laundry soap, I rehearsed the alphabet from *dit-dah* to *dah-dah-dit-dit* and back again. When the twenty-six letters, ten numbers, and several punctuation marks could be recalled at will, I attacked the pages of the *BJM* and my course books, converting the printed lines into patterns of *dits* and *dahs* that I repeated to myself and reinforced by tapping out with my fingers.

The "hambolos" among us I grew almost to dislike, more for their air of lordly superiority than for their experience and knowledge. While the other tyros and I sweated over the books, they lounged in their bunks or worked on their uniforms, bragged about the ham rigs they had built, made plans for the next weekend liberty, and speculated about what it would be like to handle a "hot circuit" in the Pacific Fleet.

Only the required dismounting and assembly of my Springfield gave me any pleasure on those all-too-short August and September evenings. When it came to small arms, I was clearly superior to most of my classmates, for I had grown up around guns and killed my first buck at the age of fourteen, in a precipitous canyon on the American River's North Fork. I could hardly wait for our scheduled trip to the marine rifle range.

"How come you never learned the code in the reserves, Mason?" one particularly condescending ham asked me once when I was rubbing down the bolt mechanism of my rifle with a lightly oiled rag.

"We had only one guy who could have taught us, and no equipment," I explained.

"Then what the hell did you guys do?" he demanded.

"We drilled some," I said. "We dug our own rifle range with the help of an ex-marine whose kid is a V-3. We got a bunch of rifles from the reserve armory in Sacramento, and we qualified on the range."

He looked at me as if I were some kind of country bumpkin. "You're supposed to have learned the code. This is the communication reserve, not the infantry."

"And you're supposed to have learned how to drill and shoot a rifle," I said, starting to get angry. "Did you ever read about your military duties in *The Bluejackets' Manual?*"

He laughed scornfully. "Frig that military shit. The navy wants me for what I know—and that, my friend, is *ray-dee-oh.* You'd better concentrate on the code, Mason. Or you're going to end up bringing me coffee."

I fitted the bolt into the well of the receiver and slammed it home. Then I raised the piece to the port arms position. "And you'd better concentrate on shoving off, buddy, or I'm gonna jam this barrel right up your *culo.*"

He did, with alacrity. He was right about the code, but how could I have learned it if the navy hadn't given our small V-3 detachment so much as a practice oscillator? Now, the question was academic. Where I was going, I had been told, excuses of any sort were not tolerated. I had better know it, period!

To Craighead's great relief, it was not long before we novices had made enough progress to be shifted from copying his hand transmissions to the automatic code machine. Operating on punched tape, it could transmit perfectly formed characters (one unit for dots; three for dashes; three for the pause between letters; six for the pause between words) at any speed. The disadvantage for us was that we had to switch from transcribing by pencil and pad to the typewriter. I thanked God for two blessings: that I had taken a full year of typing as a senior at El Dorado Union High School and had reached a respectable speed of around fifty wpm with few errors; and that I was using the same trusty Underwood model that I had learned on. The only difference was that the navy radio mills made capital letters only, eliminating the tedious raising and lowering of the carriage.

No sooner had we worked up to copying eight or ten five-letter code groups per minute than a new challenge was presented: learning to transmit with the telegraph key. This was more difficult than it seemed, since it required a 180-degree reorientation. In copying, what the ear receives in aural symbols the brain must translate into letters (or numbers) that the hand, reinforced by the eye, can record on paper. But sending begins with the eye, not the ear. The message is converted by the brain in reverse, from sight to abstract Morse equivalent. It must then be forwarded by the hand and wrist, working in harmony with the eye and brain, the feedback sound in the earphones providing assurance that the retransmission is an accurate one.

Even when the eye, brain, and ear are working perfectly, the hand and wrist may fail. If surgeons are born, not made, so too are radio operators. The kind of coordination involved is similar. No matter what the brain may know, if the wrist and fingers cannot execute its commands, one can only harm patients—or send gibberish. Later, I knew many operators who could copy plain language at speeds of fifty or sixty wpm, but only a few whose skill at the key was comparable. (One of these was Joe Goveia, an ex–merchant marine who served with me in the *California* and *Pennsylvania*. He could send twenty-five wpm all day with a precision and timing that approached those of a machine. Despite all his advice and coaching, I could never rival his proficiency. God had given him "the wrist and the fist."

Fisher, my hypercritical and very bright buddy, was having an even worse struggle than I. His eye-hand coordination was not good to begin with. While he had been a straight-A student at high school and a whiz at math, he had trouble stopping a softball in our gym-class games and always played far back in the outfield. Adding immeasurably to his problems, he was left-handed. All the telegraph keys were recessed on the right-hand side of the typewriters. This meant that Fisher had to lean far to the right, his left elbow pressed awkwardly against the mill, to even reach the button. He was the only portside radio operator I ever saw; the fact that he was not washed out at this stage is a tribute to his dogged Germanic determination.

At the time, I must admit, I rather enjoyed his discomfiture, knowing quite well he would not fail.

At last the time came to actually fire the Springfields we had been drilling with for weeks on the Grinder. We were taken by bus to the marine-corps rifle range at nearby La Jolla, where we would fire five rounds each in the prone, sitting, off-hand, and prone rapid-fire positions. Just as we V-3s had been allowed to skip the detention barracks, so we bypassed the .22-caliber range at the training station, used for indoctrination of regular-navy boots before they were trusted on a real range. It was assumed that we already knew range procedures. I couldn't refrain from chuckling as I surveyed my edgy classmates in undress whites and leggings, the brims of their white hats pulled down over their foreheads for protection from the glaring sun. It promised to be an interesting day.

Rifle ranges, military or civilian, have always held a certain fascination for me. Being able to hit a miniscule round black target at 200 or more yards with a tiny projectile shot from a long steel tube at double or triple the speed of sound, after making the proper adjustments for windage and elevation, is no mean accomplishment.

Then, there was the special atmosphere of a rifle range: the men were alert, careful, disciplined. One was aware that any man carrying that simple yet marvelously efficient steel tube, when armed with its cupro-nickel-jacketed cartridges, could easily kill any other man; the knowledge made one speak a little more softly, move a little more carefully, and be doubly attentive to the strictly orchestrated proceedings.

Our chief and bosun's mate first were both there in dungarees, conferring with the range master, a marine top sergeant who stood six feet two inches without an ounce of fat and whose sun-coarsened face was all straight lines and angular planes. The profession of arms was his milieu, and there was no doubt of his thorough competence as he explained the first course: five rounds slow-fire from the prone position.

"All right, men. Load and lock."

I drew the bolt back, pressed the clip down into the magazine with my thumb, closed the bolt, and flipped the safety lock up.

The range master looked up and down the line. He was taking his time with the reserves. I saw nervous V-3s on both sides of me. "Nothing to it," I encouraged them. "Remember two things. Don't put your thumb across the stock. And don't yank the trigger. Squeeze 'em off."

"Ready on the right," the marine said, following the traditional and oddly poetic ritual. "Ready on the left. All ready on the firing line."

There was a pause while all hands eyed the down-range pits where the targets were concealed. "The flag is up," we heard. "The flag is waving. The flag is down."

The target frames began to emerge slowly from the pits. Prone, we waited tensely.

"Commence firing!"

A ragged volley followed immediately. I took the time to carefully center the front sight in the rear-sight notch and zero in on the lower quadrant of the bull's-eye before squeezing the trigger. I tried to do this so gently and with such concentration that the actual firing would take me by surprise. I also mentally noted where I had been aiming at that precise instant. Not because the *BJM* recommended it as a cure for flinching—I wanted to know if the rifle was shooting where I pointed it, so I could make any necessary adjustments.

After the allotted time, "Cease fire!" was given and the targets were hauled down. When they came up again with the markers, our stone-eyed chief ran down the line with his binoculars, shaking his head slightly at what he saw. On my left, the ham who had advised me to concentrate on the code was sucking on a bloody thumb.

"Read any of that 'military shit' in *The Bluejackets' Manual* lately?" I shouted across. "Maybe you ought to check page 503." I smiled when I said it, but got no smile in return.

The chief came over. "Mason?" he asked, not really sure of my name.

"Yes, sir."

"Where'd you learn to shoot like that?"

"I grew up in the Sierra foothills," I said. "I've been hunting and shooting since I was eight or nine."

He eyed me with new regard. "Maybe you should be a gunner's mate instead of one of these *dih-dih-dah* people," he said.

It was the first praise I had received from the navy—and from an unexpected source. "Maybe I should," I agreed, thinking of the infernal code machine. "Thanks, chief."

He turned to the V-3s around us. "When you can shoot and drill like Mason here," he said in his hard, brittle voice, "you just might be ready for the fleet."


That was a rare moment of triumph. Mostly I was in the dunce category along with the other nonhams. Weekend liberties following personnel inspection by the commanding officer, Captain H. C. Gearing, offered some relief from the pressure of performing better than I knew how. But not much.

While the older V-3s stayed on the base and wrote letters or went to the movies, which were screened in a large open-air enclosure between two barracks, Fisher and I wandered along Broadway in downtown San Diego. Usually we were accompanied by a new friend, R. S. (Dick) Stammerjohn,

Recruits in dress blues leave the main gate of the San Diego Naval Training Station on weekend liberty. The photo was probably taken ten or twelve years before the author's radio-school period. Little but the autos had changed. The gate is still in use today as a service entrance. (Photo courtesy of the San Diego Naval Training Center.)

who at eighteen or nineteen was already a skilled ham operator and a cinch to make third class.

The first time I saw Dick in the shower, though, I wondered how he had ever passed the navy physical. His blond, wedge-shaped head was incongruously large in proportion to his bony, emaciated physique. He looked so frail that I would have worried about his safety in a moderate gale. But this appearance was most deceiving. A sharp intelligence shone in his eyes, a flow of uninhibited, polysyllabic words fell from his lips, and a great *joie de vivre* animated his 5 feet 9 inches and 120 pounds.

But not even Stammerjohn's élan could prevail against the cruel reality of San Diego: we were boots in a city full of boots, and we were too young to buy a drink, as real sailors did. All too often, we found ourselves at the navy YMCA, along with hundreds of other navy and marine-corps recruits.

The six-story "Y," occupying an entire block along lower Broadway, offered a multitude of activities, mostly free and none costing more than a dollar. All of them left me either indifferent or bored. I didn't care for swimming in teeming basement pools or watching basketball teams with which I had no personal identification. I hadn't joined the navy to sing "old favorites" around the piano in the lobby or to listen to Jack Benny's Sunday radio hour in the library. (Which speaks more directly of my own idiosyncrasies, of course, than of the earnest efforts of YMCA officials to provide clean, wholesome outlets for young men far from friends and family.)

A couple of times I took advantage of the free transportation offered by the "Y" to attend a Christian Science church. The driver of the church party was a retired army colonel, a tall, austere man who gave the vague impression, intended or not, that he was doing us few of that faith a favor.

In this fine period photo from the World War II era, three forlorn boots from the naval training station face the same cruel reality that the author and his buddies did in 1940. They are too young to buy a drink at the Bomber Cafe. (Photo from the San Diego Historical Society—Title Insurance & Trust Collection.)

After being largely ignored by the churchgoers, I decided that he was not, and stopped going.

Downtown San Diego was dirty and sleazy. For a dozen blocks east from the Broadway and navy piers, the merchants devoted themselves to the welfare of sailors and marines. There were many bars like Bradley's 5 & 10, and a number of cabarets—the Paris Inn, the Rainbow Gardens, and The Hofbrau—all guarded by husky bouncers who checked IDs closely and barred all who were not twenty-one. Approaching these in number were glass-fronted coffee shops like the Silver Dollar, serving thin greasy hamburgers and fat greasy French fries in the hot glare of florescent lamps. The most famous of these short-order places was the Right Spot, which specialized in ham and eggs grilled in the front window. There a crowd usually gathered, marveling at an ambidextrous cook who manipulated three smoking skillets, flipping ham, eggs, and hash-brown potatoes with controlled abandon and flair. One had to wait for a stool or booth at the Right Spot.

In those pre-television days movies were the great escape medium. At theaters with names like the Aztec, Egyptian, Mission, and Fox California, there was film fare for any taste and cultural level—from Greta Garbo in *Ninotchka* through *Northwest Passage* (Spencer Tracy) to Kay Kyser in *That's Right You're Wrong*. On Third Avenue, the Hollywood Burlesque Theatre advertised "seven big comedy scenes" and "nine big musical numbers" featuring Tracy Lamarr and Helen Jean. "Amateurs wanted," the marquee sign announced cryptically.

Nourishment for the outer man was not neglected. On nearly every block could be found a "locker club" equipped with rows of green metal lockers and minimum, usually dirty, toilet facilities. Here enlisted men could shed their uniforms and change into the "civvies" that enabled them to temporarily escape the onus of being sailors in a navy town. This type of establishment was warned against in sharp tones by the *BJM*.

Lockers rented for about $1.50 a month, and it was customary for two or three sailors to split the expense. For the several thousand men in the "tin-can" squadrons at the destroyer base at the foot of Thirty-second Street, it was a sensible investment. Not so for the recruits, who chanced losing their civvies when they were shipped out. An even better reason for foregoing locker clubs was given many years later by my friend Tom Hall, a former aviation radioman and fellow member of the Pearl Harbor Survivors Association. "Who the hell did we think we were fooling with our GI haircuts?" he laughed.

Hard by the locker clubs, if not part of them, were the tailor shops that specialized in custom-made uniforms for officers and men. The *BJM*, which viewed these with even more disapproval, was right about one thing. "Tailor-mades" were expensive. A good dress-blue uniform in a hard-

In April 1943, when this photo of navy and marine recruits was taken, the Hollywood Burlesque Theatre was no longer advertising "amateurs wanted" on its marquee. Undoubtedly, most of the apprentice ecdysiasts were by then gainfully employed in defense industries. (Photo from the San Diego Historical Society—Title Insurance & Trust Collection.)

The proprietor of this San Diego establishment obviously believed in advertising. Just as obviously, demand for his services was great among the recruits at the naval training station. In this wartime photo, dated 10 April 1943, the emphasis has changed from tailor-mades to repairs and alterations of regulation blues. (Photo from the San Diego Historical Society—Title Insurance & Trust Collection.)

finished gabardine with real flaring bell bottoms, lots of fancy stitching, and a silk inner lining would cost from fifty to seventy-five dollars—a month's pay or more for a typical seaman first or petty officer. Nevertheless, many sailors wanted them, precisely because the cloth and cut were not regulation. Such a uniform immediately identified its wearer as a man who knew his way around the navy. And a brave sight he was, rolling slightly as he navigated streets he obviously owned, white hat pitched jauntily at the back of his head, cuffs turned to display the embroidered lining, the neckerchief pressed flat (never rolled, as the regulations specified)—there went the true professional sailor! I could hardly wait until I got my own set of tailor-mades.

Approaching the locker clubs and tailor shops in great number were the credit jewelers, who dedicated themselves to the proposition that every enlisted man should invest heavily in wristwatches, rings, and cigarette lighters for himself and in diamonds for his girl friend, fiancée, or wife. Oddly, those stern moralists who compiled the BJM did not warn against dealing with these entrepreneurs. Even for an apprentice seaman, it was surprisingly easy to get credit: a glittering gold watch could be yours for thirty, forty, or fifty dollars, on easy terms of twenty percent down and monthly payments of three to five dollars. So it was that I acquired a seventeen-jewel Gruen with a gold mesh band for $47.95 (plus credit charges). This was more than I had paid for a 1929 Chevrolet roadster in running condition a year before. The fact that I would shortly be departing the naval training station bothered the smiling proprietor not at all; he merely required my name, rate, service number, current address, and my parents' address, along with the down payment. I learned later, from my shipmates who were not as diligent as I in sending a monthly money order, that the navy made sure one met one's contractual obligations.

Opposite the jewelry store was grimy Horton's Plaza, a half-square-block refuge for the unemployed and the unemployable, casualties of the Great Depression who sat sullenly or soddenly on the long benches staring at defeat. The presence of sailors brought a few shuffling over, heads down, to beg "a dime fer acuppa coffee." The picture of free Americans reduced to panhandling was profoundly upsetting; my inability to do anything about it, beyond an embarrassed exchange of small coins, made me feel vaguely guilty as well.

At the zoo in Balboa Park there were few beaten faces, and the only merchants were purveyors of popcorn, cotton candy, Orange Crush, and Delaware Punch. But the animals made me homesick for my dogs and my gun. And the giggling high-school girls in their pleated skirts, colorful blouses, tan-and-white saddle shoes, and bobby-socks reminded me achingly of the girl I had lost and could never have again. I watched the bolder of my fellow recruits approach them and invariably be rejected, and saw no reason to believe I would fare any better.

Usually, we were back at the barracks long before midnight. Already, we were acquiring the inevitable part of a sailor's education that bred mistrust, bordering on dislike, of civilians.

In late August, the annual Navy Relief Carnival provided a welcome break in our routine. Fisher and I joined a mob of some twenty thousand people at the adjoining marine base, where we rode the Ferris wheel and the "loop-the-loop"; watched a daredevil navigate a motorcycle around the walls of a tall vertical cylinder at high speed; tried our skill at the shooting gallery (where I won a couple of celluloid Kewpie dolls with painted pink faces that I promptly gave to a sloe-eyed senorita); and ate quantities of hot dogs and drank gallons of coffee and Coke. By 2300 we had pushed near the front of a temporary stage for the *pièce de résistance*: the appearance of Barbara Stanwyck and Robert Taylor from Metro-Goldwyn-Mayer Studios.

The only movie star I had ever seen was Randolph Scott, the tall, leather-faced Western hero, who appeared one day in Placerville driving a magnificent front-wheel drive Cord cabriolet, so I awaited the entrance of Stanwyck and Taylor with great interest. The former was attractive, her red hair in a fashionable high pompadour, but she was a trifle long of nose and spare of figure for my taste. She didn't begin to compare to my beloved, who was just the opposite—an opinion that drew a contemptuous laugh from Fisher. But Taylor was as handsome as the cameras made him appear. He gave an earnest little speech about the importance of what we navy and marine recruits were doing for our country in its hour of danger. It did not seem a sham and I put him down as a regular guy, an opinion I never found reason to change.

Three weeks later, as I was preparing for exams in what I felt was a losing race against my enemy, Time, we received an unexpected visit from a civilian of awesome rank: Secretary of the Navy Frank Knox. He had just returned by Consolidated flying boat from Pearl Harbor and consultations with Admiral James O. Richardson, commander in chief of the U.S. Fleet.

Since his one-day agenda called for dinner with the recruits, an elaborate four-page menu was printed in blue and gold on a pale cream stock, and souvenir copies were made available to all. Reading it today, one is struck by the amount of useful, even historically valuable, information it contained. It devoted one page, for example, to the bill of fare for an entire week at the training station mess halls. While a nutritionist of the 1980s might consider our diet to have been somewhat heavy in fats and starches and, consequently, a menace to the waistline and the cholesterol count, there is no doubt that it was both ample and nourishing. Most of us, in fact, had never eaten so well in our lives. Exploring the fine print, one is told that an estimated total of 23,800 rations would be served for the week beginning 16 September at a ration cost per day of about thirty-nine cents. Only thirty-nine cents to feed a voracious, growing eighteen-year-old three meals!

The menu that noon was a typical one for Tuesdays: chicken rice soup; fried chicken with giblet gravy; mashed potatoes; buttered peas; tomato and lettuce salad; rolls and butter; loaf cake with ice cream; and a choice of lemonade or coffee.

Being excused from the review and dress parade staged by the recruit regiment for Knox's inspection, we V-3s got into the mess line early for a close-up of our leader. Many years later I read with a certain dismay that Colonel Knox, one of Teddy Roosevelt's Rough Riders in the Spanish-American War, knew nothing about the navy, confiding at the time of his appointment that he was excited about "playing sailor."* But on 17

*A.A. Hoehling, *The Week Before Pearl Harbor* (New York: W.W. Norton & Co., 1963), p. 31.

September 1940 we did not know that he lacked both experience and influence in Washington. Certainly his appearance in our mess hall, where he was escorted with the utmost solicitude and deference by two rear admirals, two captains, and a gaggle of civilians, seemed to confirm his importance.

I had entertained a vague hope that I might find myself next to the secretary in the chow line, where we could exchange a few pleasantries. Perhaps he would ask me if there were anything he could do to ease my burden, and I had a suggestion or two firmly in mind. I could visualize the new respect with which I would be treated by Craighead, the chief, and the first class who supervised with curled lip our clumsy evolutions on the Grinder. But nothing of the sort happened. We were already through the line and seated when Knox and his party got into it with the self-conscious jocularity of men who are trying to impress their inferiors.

Between bites, we stared at Knox and his cohorts, seated at a nearby mess table. The secretary did not look very impressive. He was a short, lumpish man in a double-breasted gray business suit, with a lot of sandy-gray hair and heavy-lidded, slitted eyes above fat cheeks and jowls. He was a wealthy Chicago newspaper publisher turned politician, but he loomed larger than life then. He was the first famous man I had seen at close range (movie stars hardly counted). He had been Alfred M. Landon's vice-presidential running mate in the 1936 election. Now he was the top official of an entire navy, of which I was such an insignificant part. And above all, he was a close associate and advisor of President Roosevelt. He actually talked with Roosevelt! Now he was in the same mess hall with me, not twenty-five feet away, eating from a mess tray I possibly had pushed through the line a day or so before. The emanations of power hovered around him like an aura, and what he symbolized seemed more real than what he was. I was glad, despite the psychic distance, that I shared the same mess hall with him, and gladder still that I was a member of an organization that made it possible to shine in such reflected glory.

My Gethsemane came three days later, when we assembled in the classroom for the proficiency tests in theory and code. The written examinations arrived in sealed envelopes from the Bureau of Navigation in Washington, eliminating any possibility of cheating or collusion. "You're a natural exam taker," I had been told in school, and never was this facility more welcome than on 20 September 1940.

I used every second of the allotted time before handing over the log-size sheets. Next we went to the typewriters to copy five-letter code groups at the required eighteen-wpm speed. Soon I was seated opposite Craighead at the telegraph key for the last and, I had assumed, most critical part of the exams. He noticed my nervousness and smiled.

"Relax, Mason," he said. "If you pass everything else, I'm not going to bust you for your fist. You've worked like hell and you're going to be all right. Just keep on working."

When the test results were posted early the next week, there was a rush for the bulletin board. I was dancing up and down like a boxer in my impatience to learn my fate. Fisher and I finally got close enough to scan the alphabetical list. We had both passed!

We whistled soundlessly and shook our heads in unison and relief and clasped hands. Then we found Villafranca and the other neophytes on the pass list and congratulated them. Stammerjohn and the other hams got more perfunctory salutes, since their success was a foregone conclusion. We all laughed and joked and hugged each other as if we had just won an important football game against all odds. And we found time to commiserate with those who had failed and would have to go to the fleet as seaman-second strikers.

My final mark in the training course for radioman third class was 3.2: better than I had expected and better, undoubtedly, than I deserved. In the "A to N" course, my mark jumped to 3.73, reflecting my chief's evaluation of my performances on the Grinder and at the rifle range.

This being the first class of V-3s to pass through the training station, we were honored at the brief graduation ceremony by the presence of both the commanding officer and his exec, Commander Marion C. Erwin. Captain Gearing, looking competent and more than a little forbidding, presented us with our BUNAV training course certificates. The station photographer was called in for group photos. When we got the prints there was a great signing of names, home towns, and ham call signs on the backs. Through the good offices of Chief Radio Electrician Charles A. Mattson, our division officer, the commanding officer, and the exec signed them, too, on the front. I was proud of this photo and managed to keep it safely through all the vicissitudes of war and peace.

Many of the great decisions of life are made with a casualness that belies their serious and possibly somber consequences. In the South Pacific in 1944, I cut cards with the other first-class radioman in the *Pawnee* (ATF 74) to decide who would return to the States for radio school after nearly two years in and around the Solomons. I turned up a nine, and my junior, a jack. I shrugged as if it did not matter and strode smartly from the wardroom, conscious of the appraising and, I hoped, admiring stares of the skipper and the exec. As they well knew, it did matter: ahead were the invasions of the Palaus and the Philippines and "Crip Div 1," when we were used as bait by Admiral Halsey to lure out the Japanese Home Fleet while towing the *Houston* (CL 81) from Formosa to Ulithi under air attack. I did not get a scratch. Yet, a thing as random as cutting a card from a deck had decided whether I went home or remained in danger.

Graduation ceremonies for the communication reserves, San Diego Naval Training Station, September 1940. *Seated, left to right*: Commander Marion C. Erwin, Captain Henry C. Gearing, Chief Radio Electrician Charles A. Mattson. *Front row, standing*: unidentified chief boatswain's mate, Floyd C. Fisher, the author, Floyd J. Slosson, Robert H. Potts, Jack Villafranca, unidentified chief quartermaster. *Second row*: R. S. (Dick) Stammerjohn, Cleo C. Mooney, Kenneth Cochrane, Lloyd D. Wingert, Robert L. Hobson, William C. Daly, R. E. Rainie. *Top row*: Ben Ritter, Stephen N. Wyckoff, Bill DeArmond, and Jim Jett.

On a morning in late September of 1940, I made a decision about the ship I would serve in nearly as casually as if I had let the cards decide, and its effects were more far-reaching still. After receiving the good news that I had passed the exams and would receive my rating as RM3, I and the rest of the V-3s assembled in the classroom that had so nearly been our nemesis. The atmosphere was decidedly more relaxed: we were going to the fleet. I felt grateful to Craighead and thanked him for his instruction and encouragement, as did a number of others. In rare good humor, he repeated his counsel of examination day. I realized, even then, that getting so many of us through the course was a remarkable performance; he had passed his test, too.

We were told we could have our choice of the battleships listed on the blackboard. All of the storied names of the battle line were there, including the *Arizona, Oklahoma, Nevada, West Virginia*, and *California*. The list was like a roll call of the States—it meant little more than that to me. The older V-3s, who had been on summer cruises, knew which battleships were good duty but, perhaps wisely and certainly selfishly, kept their counsel. As for the rest of us, we had heard that the home port of the battleships was San Pedro, but didn't know where any of them were at present. We didn't know that the *Pennsylvania* was the flagship of the admiral commanding the U.S. Fleet (CinCUS); that the *California* hosted her own four-star admiral, Commander Battle Force (ComBatFor); that the *West Virginia* flew the three-star flag of Commander Battleships, Battle Force (ComBat-Ships); and that the *Arizona* was Commander Battleship Division One (ComBatDiv 1), with a two-star admiral on board. We didn't know that the *Tennessee* was considered a "happy ship," as was the *Maryland*; that the *New Mexico* was relief flagship for CinCUS and ComBatFor and therefore to be avoided; and that the *Colorado* was the "eight ball" ship of the line. All this, and much more, was known to Craighead and Mattson but they, quite understandably, did not tell us. If they had, it might have been difficult to explain why the entire V-3 complement asked for duty in the *Tennessee* and *Maryland* (as several senior V-3s, in fact, did).

Fisher and Stammerjohn and I were sitting next to each other. We stared at the blackboard in perplexity.

"Well, Mason," Fisher demanded. "What ya think?"

I thought about it for a moment or two. Arizona was cactus and sage-brush, but it had Zane Grey and the Grand Canyon and the Mogollon Rim to recommend it. Oklahoma was the birthplace of the woman I loved, but she was now the mother of another's child and I did not need this constant reminder. Nevada was Arizona minus the scenery. West Virginia was coal mines and hillbillies. California was, of course, my native state. Tennessee was moonshine whiskey and twanging guitars. Maryland was Annapolis and the U.S. Naval Academy—I paused over that. Pennsylvania was Ben Franklin and the Liberty Bell and the *Saturday Evening Post*, and, like Maryland, had to be considered. New Mexico was Indian pueblos and the Santa Fe Trail, but its arid deserts were doubtless overrun by Mexicans trying to reclaim the territory we had taken from them in 1848. Idaho was fine if you liked to hunt bighorns in the snow. Mississippi was "Old Black Joe" and cotton plantations that looked like Tara in *Gone With The Wind*. And Colorado had no mountains to compare to my own "range of lights," the Sierra Nevada.

"Well," I said, drawing out the word for a little more study of the board. "The *California?*"

My buddy nodded vigorously. "Yeah. The *California*."

"Me too," Stammerjohn chimed in. "I don't see any battleship named the *Washington* around here."

That is why I was not serving in the *Arizona* or *Oklahoma* or *Maryland* or *Pennsylvania* at Pearl Harbor. If I had chosen the *Arizona* or *Oklahoma* how different life would have been for those near and dear to me, past and present.

And how different for me.

CHAPTER 3

# Destination Diamond Head

"Now man the special sea detail!"

The flat, authoritative voice from the bridge, amplified over the "squawk box," sent seamen scurrying to the forecastle deck of the oiler *Neosho* (AO 23). And sent tingles racing up and down my spine as I stood with Fisher and Stammerjohn at the starboard rail of the after deckhouse. We were going to sea for the first time in our lives—destination Hawaii!

It was 1030 hours on the rare cloudless morning of 6 October 1940. The Prep—a dark blue flag with a square white center—snapped from a signal halyard abaft the bridge, announcing to all ships in San Diego Bay our intention of getting under way. The ship's siren sounded off with two loud *whoops* ending on urgent rising notes. The deep-throated steam whistle followed with a long, falling *woooo* that reverberated across the harbor. Now the starboard anchor was weighed to a shouting of orders, a rattling of chains, and a whining of electric windlass. Almost imperceptibly, the *Neosho* began moving ahead.

Wordlessly, we watched the tanker gain headway past the docks of downtown San Diego. The city, spread across its eastern hills, looked scrubbed and glowed in the bright sunlight. Balboa Park was a square-cut emerald between the bay and the heights. Even the grimy commercial area acquired a new respectability as I traced the arrow-straight progress of Broadway by its tall financial and real-estate buildings. The naval training station was soon off the starboard quarter, and I could see geometric formations, indistinctive masses of white at this range, drilling on the Grinder.

It seemed a long time ago that my two friends and I—the only ones to choose the *California*—had packed our seabags, lashed our hammocks around them in the approved fashion, and been taken by motor launch to the *Neosho*, anchored off the North Island Naval Air Station among the four-stacker scout light cruisers. Actually, it had been only yesterday. With the profound relief of an Odysseus who has been to Hades and back and finds himself safely en route to Ithaca, I bade a silent farewell to San Diego

Naval Training Station. Looking ahead, I could see the long, narrow peninsula of Point Loma, the old light station commanding its heights, and the ocean beyond, glittering with a million facets of reflected light.

We cleared the jutting spur of Ballast Point, put the coast-guard lighthouse, a thin cylindrical tower at Point Loma's very tip, on our beam, passed two sea buoys, and met the long, slow northwest swell of the Pacific.

Immediately, the *Neosho* started to roll. We changed course to west-southwest, aiming for Hawaii, and increased speed. The seas began to come in nearly on our beam. The *Neosho* reacted to them in ponderous motion, like an elephantine motorcyclist maintaining his balance on an undulating road. Clinging to the rail, I looked back at the fast receding and now puny skyline of San Diego—a city I would always remember but never miss—and ahead to the far reaches of blue ocean continually meeting a lighter blue sky.

"Isn't it great?" I shouted to my companions.

Stammerjohn, smiling broadly, gave a thumb-up gesture of approval. But Fisher, his freckled face even paler than usual, managed only a sickly green grin.

After we had been assigned a lifeboat at abandon-ship drill, we reported to the radio shack, which we had found on our exploration of the ship the previous day. Strategically located one deck below the navigation bridge in the midships house, it was a large, airy compartment equipped with several operating positions, a couple of bulky transmitters, a built-in desk for the supervisor, and the inevitable electric coffee pot.

To my dismay, Stammerjohn had immediately volunteered our watch-standing services to the radioman first class in charge. The latter looked up from his radio logs long enough to discover we were V-3s fresh from radio school. Tersely, he instructed us to report to him the next afternoon. I needn't have worried about being given an assignment beyond my capability. As it turned out, all Stammerjohn had done was introduce us to a continuing, six-day cleaning detail. We thereby learned the truth of the old navy axiom, "Don't *never* volunteer for *nothin'*."

Although the radio shack looked clean enough to pass material inspection, we soon found ourselves swabbing and waxing the red linoleum on the deck, cleaning the encrusted salt from the thick glass of the ports, and polishing all the brightwork both inside and on the wide, cambered deck outside.

We might have been newly hatched third-class radiomen to the Bureau of Navigation, but to our supervisor we were strikers—and reserve strikers, at that. (In navy parlance, a striker was a seaman in training for a specific third-class petty-officer rating.) The distinct impression he gave was that we were a very low form of life and so lacking in intelligence that we could not be trusted even with the simplest detail.

The USS *Neosho* in 1939 peacetime garb and riding light, just as she was on the author's first sea voyage from San Diego to Pearl Harbor in October 1940. The radio shack was one deck below the navigation bridge, port side. The *Neosho* was lost on 7 May 1942, during the Battle of the Coral Sea. (Photo courtesy of the National Archives.)

The first class had a John Doe face and a grating personality. I remember him standing over me and sharply calling my attention to some minute corner of deck that remained unpolished or a dog on a port that did not gleam with the required patina. Thereafter, I showed such devotion to my duties that I put extra-heavy coatings of wax around his desk. Perhaps he would take a pratfall right before my innocent eyes and bruise his coccyx. But he was as quick as a cat and as suspicious as an insurance adjustor. After a couple of slips and slides, he ordered me to take up the extra wax and repolish the entire deck.

Following an impassioned plea by Stammerjohn, he reluctantly gave us permission to sit in on the NPM Fox schedule broadcast from Honolulu as backup to the regular watch standers. This made our dedication with swab, squeegee, and rag worthwhile, for it was a preview of what I expected to be facing soon in the *California*. The broadcast was transmitted at my top receiving speed. It was considerably more difficult to copy than the San Diego code machine because of fading and static caused by atmospheric conditions. But each code group was sent twice, so any letters that were missed could be filled in on the repetition.

Even with this schedule, I found time for pleasant hours on the fantail, drinking coffee and watching the boiling white wake of the oiler, as broad and straight as a superhighway, dissipate toward the eastern horizon. Or I took the catwalk forward, through the bridge house to the bows, where I could watch the two creamy waves the stem created as it parted the water—rather like Paul Bunyan plowing virgin soil with his great Blue Ox, I fancied.

We seemed absolutely alone on a boundless ocean. Feeling the wind in my face and hearing the splash of the bisected sea, I gloried in our isolation from man. Here it was clean, bracing, serene, especially at night when the ship behind me was blacked out except for her red and green running lights, and masthead and range lights. I knew at such moments that I had chosen the right service. But I was aware that the sea could be an implacable foe as well, that it was menacing and unpredictable not only on its surface but fathom by fathom to its unthinkable depths; so my affection for its beauty was tempered with a healthy respect. That observation I had learned a few days before to apply to women as well.

The members of our V-3 class had chosen half a dozen battleships for duty and were soon departing, singly or in small groups, for San Pedro or Bremerton or Lahaina Roads or Pearl Harbor.

But before we all scattered, to wherever destiny and the U.S. Navy would take us, there were certain graduation rites to be observed. Not all of the radiomen participated, by any means, but the seven or eight of us who were both young and "gung-ho" did. Having worked very hard, we had passed the first and most critical test to receive the first reward: the silver eagle, stylized lightning bolts, and single red chevron of radiomen third class. That event was worthy of celebration.

The second reward quickly followed: a seventy-two-hour liberty. We had been paid and our billfolds were thick with ones and fives as we headed for the main gate. I remember walking with left arm slightly akimbo for the benefit of the pitiful boots with the shaven heads, white shoulder braids, and single cuff stripes of apprentice seamen.

As befitted men wearing "crows," we enjoyed the luxuries of a cab ride to the YMCA (twenty-five cents each) and double-occupancy adjoining rooms overlooking Broadway (seventy-five cents). Now the true grail could be pursued: bourbon whiskey. Like all rewards for purity, there were challenges to be overcome, the first of which was the fussy Protestants who ran the "Y." Not only did they disapprove of alcohol, but they gave the navy full cooperation in protecting the morals of the young servicemen entrusted to them: neither liquor nor girls were allowed in their celibate rooms.

The next hazard, and a more difficult one, was the liquor-store proprietors. These bleak-eyed fellows would happily have sold us anything we could afford to buy, regardless of age, but were constrained by their fear of the state liquor board and demanded identification of enlisted men without hash marks before relinquishing the prize. In any group of sailors, however, was one knight whose ID proclaimed, falsely or otherwise, that he was twenty-one. We had such a man, a tall, thin Southerner, who took our pooled resources and soon returned to the YMCA with several half-pints and pints cleverly distributed about his person.

Using the squatty bathroom tumblers, some of us took it straight and some diluted with warm Coke. There we were, peers all, bound not only by a shared accomplishment but by the common uniform we wore, drinking to the immediate past and to an unknown but, we were convinced, adventurous future.

Soon our boisterous conversation turned toward the opposite sex and the fact that we new third-class petty officers, conquerors of international Morse and Ohm's law, had had no success of late with those more difficult subjects. The classmate who had procured our whiskey knew what to do. He had been, he boasted, in New Orleans and Mobile and Galveston, towns with a lot more action than "Dago." He called a cab.

Tipsy, we pussyfooted across the echoing masonry of the lobby, fooling the fussy Protestants not at all, to find a battered Yellow Cab and an equally battered car jockey awaiting us. After an intense, gesticulating conversation with our leader, he opened doors and we crowded in. A drive as short as the negotiations brought us to a dark doorway on a dark street. It did not seem like a good place to be, but the veteran of Mobile and Galveston and New Orleans was undaunted. Paying the cabby off with a few dollar bills, he led us on a charge up the stairs.

We found ourselves in the reception room of the first "cat house" I had ever seen. The furniture was overstuffed, hard, and covered in the dark velour popular in the 1930s. The light from brass-stemmed floor lamps was dim. The wallpaper was brown and patterned, background for ornately framed prints of voluptuous nudes. The air was close and musty, as if the windows were never opened, the heavy drapes never drawn.

The women came in and fell professionally upon my friends. Since there was little emotion involved, my buddies would be little affected and would joke about it the next day in their regimented but secure quarters. For the women, the collectors of their grubby slips of paper, there was no such retreat.

I must have shrunk back against the arm of the massive sofa, thinking these thoughts, while the selections were being made. I felt a little sad and sober and afraid.

But I was not alone for long. A tall dark-haired woman in a modest, flowered housecoat came in. She seemed sophisticated and older, around twenty-five, and did not even look like a prostitute.

"Hi!" she said, smiling. "My name is Marie. May I sit down?"

She did not sit too close. "Tell me about yourself, Sparks," she said in a low, soft voice. "Is that a brand-new crow?"

Working in a navy town, naturally she would know about such things, I reasoned. Nonetheless, it was reassuring. I couldn't very well brag to a friend like Dick Stammerjohn. He knew that if I hadn't been given the benefit of the doubt by Craighead and Mattson, I would be going to the fleet as a seaman second. But I could talk about my achievement with this slender, dark-eyed woman, who looked a little like my seventh-grade teacher. Adroitly, she led me around to the reason for my visit.

"What are you afraid of, Sparks?" she asked. "None of your friends are."

I explained, haltingly, that I had never been in such a place before. She smiled and took my hand. "Tell me something, Sparks. Have you ever been with a woman before?"

I considered the question. "Well . . ."

Now she laughed, still with me. "I know. In the back seat of a Model A Ford, with some sixteen-year-old kid."

"Seventeen," I corrected.

We both laughed. "You still haven't been with a woman, honey." She stood up, pulling me with her.

"Come along. I'm going to teach you something about them."

And she did.

I departed San Diego with my education as a sailor advanced on two fronts. To me, the bordello experience was no less a rite of passage to full manhood than the boot-camp one had been.

As I boarded the *Neosho*, my first ship, and proudly saluted the OOD and the flag at its fantail staff, I knew I would shortly have to pass one more test: seasickness. At the training station, I had heard stories about enlisted men who had to be transferred to shore duty or even discharged because they became violently ill as soon as their ships got under way. Since a sailor who couldn't go to sea was about as useful as a policeman who couldn't fire a gun, I prayed it would not happen to me.

A master-at-arms led Fisher and Stammerjohn and me clear forward, into a dim, dank, and none-too-clean forecastle. There were no ports: the only light came from widely spaced, naked electric bulbs. Ventilator pipes from the weather deck above admitted a grudging trickle of air that scarcely stirred an atmosphere fouled with a hanging blue cloud of cigarette smoke.

We had been preceded by a hundred passengers from the receiving ship, who had appropriated all the bunks near the air supply and the doors leading to the main deck. We finally found three uppers against the forward transverse bulkhead, very far from light and air. We were near the anchor windlass, and the smell immediately told us that we were also very near the head, which clearly was not maintained to navy standards of sanitation.

It was under these primitive conditions that I discovered I was one of that fortunate company who does not suffer the agonies of seasickness. Fisher did not fare so well. When we took our meals in the after house, 500 feet away and accessible only by means of the narrow catwalk over the main deck, he would merely poke at his cafeteria tray and excuse himself after a decent interval, grumbling that the chow was lousy. As for W7HEX Spokane, I don't believe he gave the possibility of any indisposition a thought.

"God, I love this salt air, Ted!" he would exclaim, shoveling in great quantities of food.

Some of our fellow voyagers were not so lucky, and their sufferings evoked clashing feelings of compassion and antipathy. One of them made the classic mistake of rushing to the windward side when his stomach rebelled against food and motion. It was not punishment enough that the wretch made a mess of his undress blues: he was spotted by a boatswain's mate who ordered him to clean up the deck with swab and bucket.

"You goddam landlubber," the boats bellowed, as the onlookers guffawed, "next time hit the lee side. Put your *ass* to the wind, not your face!"

The second day out, we moved our watches back one hour to conform to zone plus nine time. All that meant to three rated strikers was another hour with the cans of corrosive brightwork polish. We were convinced that Radioman First Class John Doe had personally arranged this with the captain to lengthen our working day.

Then we discovered the crew's reception room in the after house. It was a smallish compartment provided with narrow synthetic-leather benches secured to the bulkheads. We decided they were just wide enough to accommodate bodies prone as well as upright and got permission to sleep there. It was damnably uncomfortable; I had to sleep on my side, with my back wedged firmly against the bulkhead cushion and one hand knuckling the deck, since the metronomic motion of the ship threatened to pitch me off the bench at any moment. But I could see the stars through the open ports, breathe air that was not secondhand, and smell the salt fragrance of the sea rather than the noxious exhalations of the forecastle head.

By the fourth day the skies were leaden with stratocumulus, the gray-green sea was barely ruffled (but still not calm enough to more than damp the *Neosho*'s incessant roll), and the temperature had edged up into the mid-seventies by late morning. One of the radiomen showed me where he lived: in a roomy compartment with portholes, in the after house, that he shared with three others. The rest of the ship's company were similarly housed. By navy standards for enlisted men, this was the Waldorf Astoria, and I was more than a little envious.

In my peregrinations about the ship, I also learned that regular navy men loved to talk about ships, a subject second only to the interlocked ones of liberty, strong drink, and the opposite sex. The *Neosho*, I was informed, had a standard tonnage of 11,325, a length of 552 feet (525 at the waterline), a beam of 75 feet, and a full-load draft of 32 feet. Her twin-screw geared turbines gave her a shaft horsepower of 13,500 and a speed of a little more than 18 knots. On this run, she was carrying only about 500,000 gallons of fuel, bunker and diesel oil, and high-octane gasoline—less than 10 percent of her cargo capacity of 147,150 barrels (6,174,000 gallons). That was why she was riding light, with a foot or two of black boot-topping and a lot of red underbelly showing every time she rolled—which she did abominably. I also was told that she was a Standard Oil tanker that had been taken over by the navy the previous year, which accounted for her merchant-marine-type crew accommodations. As with most scuttlebutt, there was a germ of truth here. I later found out that the *Neosho* was a navy Maritime Commission design, one of twelve new oilers with national defense features. The *Cimarron* (AO 22), *Neosho*, and *Platte* (AO 24) joined the fleet in 1939; the other nine were initially assigned to commercial operators, but all were in naval service by the end of 1941.*

By mid-afternoon of the sixth day out, operating on zone plus ten time and in undress whites, the V-3 troika was putting the final touches on what was undoubtedly by now the most shipshape radio shack in the U.S. Navy. After a minute inspection, the first class told us that the *Neosho* was

*On 7 May 1942, during the Battle of the Coral Sea, the *Neosho* was wrongly identified as a carrier and was sunk by Japanese dive-bombers, resulting in heavy casualties. She was shortly replaced by a new *Neosho* (AO 48, ex-*Catawba*).

arriving at Pearl about 1100 the next morning, and we were excused from further duties. Rejoicing in our escape from his fastidious presence, we consigned our worn cleaning rags to the deep with appropriate ceremony. Tomorrow, fabled Hawaii!

As soon as we were dismissed from 0800 quarters for muster the next morning, we headed for the vantage point of the forecastle rails, along with many other passengers. In those days, very few Americans had ever been to the Hawaiian Islands. The United States was just emerging from the worst depression in its history. For ninety-five percent of the populace, the cost of even the five-day passage from San Francisco to Honolulu on the *Lurline* or one of her Matson Lines sisters was as far out of reach as a six-month world cruise on the *Queen Elizabeth II* would be in today's far more affluent society. The fare on the only alternative transportation, the Pan American Clipper flying boat that once or twice a week touched down at Pearl Harbor on its way to exotic Pacific and Asiatic ports of call, was even more prohibitive. Once in Oahu, a single night's lodging at one of the three luxury Waikiki Beach hotels—the Royal Hawaiian, the Moana, and the Halekulani—approximated a week's salary for a typical white-collar worker. With the exception of a few beached merchant marine sailors, vacations in Hawaii were limited to captains of industry, wealthy widows, high political figures, movie stars, and a few famous writers and journalists—unless one were in the military.

It was with great anticipation, then, that we awaited landfall. This was the only chance we would ever have, most of us thought, to see the Paradise of the Pacific for ourselves. Later, when we would return to our home towns, marry local girls, and raise families, we would be able to say we had been there. And for most of us in October of 1940, that was enough.

At about 0900, a vague mass detached itself from the azure sky off the starboard bow, deepened in hue, and assumed the shape of an explosion-wracked cone.

"Koko Head Crater," a first-class petty officer with hash marks, obviously an old Hawaiian hand, remarked. "Extinct volcano."

Most senior enlisted men ignored seamen and green third-class petty officers no older than seamen. But this one seemed a gregarious sort who would enjoy sharing his knowledge. We edged closer to him.

Dead ahead now, a long, ominous silhouette gradually sharpened into the unforgettable profile we all knew from picture postcards and South Sea movies.

"Diamond Head," we were told, quite unnecessarily.

We all gaped at the landmark, symbol of Hawaii. Photographs often lie, so that reality seldom meets expectation, but like Lake Tahoe in its tiara of peaks and the plunging headlands of Big Sur, Diamond Head did not disappoint.

There it crouched, like some titanic prehistoric beast, beetle browed and long snouted, its gaunt flanks scarred from millennia of conflict, its great claw feet immersed in ocean as if waiting to spring upon the unwary sailor in his frail craft. The impression Diamond Head gave of potential violence contrasted with the utter beauty of the morning, islands of cumulus billowing overhead, a gentle breeze scattering whitecaps over water too unearthly blue for any force but nature to paint. As that mighty mass of quenched volcano drew closer and passed to starboard, I could not but believe that it had been an object of worship and sacrifice for credulous natives. I felt uneasy, as if I were in the presence of some malevolent force I had not yet identified.

For a couple of miles west of Diamond Head, the shoreline was delineated by dense green foliage against a crescent of pink sand upon which the surf broke weakly. Occasional openings in the groves of coconut palms marked the sites of Waikiki's hotel complexes. The coral-pink one, its Moorish-Mediterranean lines looking strangely out of place in the mid-Pacific, was the world-famous Royal Hawaiian, the petty officer said. Farther along, protected behind a low offshore island, was Honolulu Harbor. At the various piers and docks, rusty merchant ships were onloading or discharging cargo. An obelisk-shaped clock tower dominated the central terminal.

Our guide pointed vaguely. "Aloha Tower. Tallest building in the islands." He smiled, adding with faint derision, "One hundred and eighty-five feet."

I said it looked a little like the one that surmounted the San Francisco Ferry Building. He lost his smile. "You won't think so after a few liberties, Sparks. Honolulu compares to Frisco like this oil barge resembles a heavy cruiser."

Behind the harbor, the city was a motley of buildings with no particular character scattered along ambling streets. Honolulu hunkered down in the lee of a range of convoluted mountains that rose up some 3,000 feet of green blending into blues and purples, fitfully illuminated with shafts of brilliant sunlight breaking through an irregular halo of clouds.

The mountains formed the Koolau Range (pronounced *ko-'oh-lau*). Compared to the Sierra Nevada, the range was little more than a chain of ambitious foothills, but it made up in sheer primitive fierceness what it lacked in foliage and scale. In its breathtaking setting between the Koolaus and the sea, Honolulu looked impermanent and inconsequential, a tangle of jackstraws carelessly flung by a giant hand.

After passing the city, the *Neosho* slowed and changed course. "Pearl Harbor coming up," we were warned. Suddenly, we were no longer sightseers on a pleasure cruise. Now it was time to put aside visions of dancing savages and their implacable gods and goddesses and consider my own rendezvous with duty and destiny. The other sailors, too, squared their white hats and stared ahead a little apprehensively as we turned directly toward the shore, passing close aboard a lone sea buoy with vertical black and white stripes. A mile beyond, a conical red buoy to starboard and a can-shaped black one to port denoted the navigable limits of the narrow entrance into the harbor. In the distance, smaller red and black channel buoys were now in sight.

We were coming into navy country, and while I did not know exactly what to expect, I had been put on notice at boot camp that it would be stern and demanding, requiring my best, most dedicated efforts. Very soon now I would join the long blue line of men who had served with John Paul Jones (as my own ancestor, Richard Dale, had done), Farragut, Dewey, and all the rest of our naval heroes. Awaiting me were friends I had not yet met, officers I had not yet served, experiences I could not imagine, in settings as remote to a country lad from Placerville as that sleepy gold-rush town was to Honolulu itself. The anticipation was awesome, it was delicious—and it was frightening. Even as I masked my anxiety, fumbling in the pocket of my jumper for a cigarette, I would not have traded places with anyone in the world.

Entering the unprotected mouth of the harbor, we rolled a little in the cross chop but soon found smooth water as we put the net tender abeam to starboard and turned right, then left again. A yard tug came alongside, and a khaki-clad figure wearing a gold-braided cap clambered nimbly up a sea ladder and headed for the bridge.

"Pilot will conn us the rest of the way," the first class explained. "You'll see why in a minute."

The *Neosho* came right ninety degrees around a sawtooth point. Pearl Harbor opened up like a cyclorama. The channel divided around a low nondescript island and grew into three estuaries dotted with navy ships of all types.

"Ford Island," we heard. "Naval air station. And there's Battleship Row."

I would have known without being told—any schoolboy of 1940 would have known. Along the near side of Ford Island, four or five huge pale-gray warships were moored in line to double concrete quays. Up close, they looked even more impressive than in the photos or newsreels we had seen. Since they were nearly bow on to us as we threaded our way between the island and the navy yard, the first thing that impressed us was their breadth. From knife-edged clipper bows, their shell platings and armored belts bulged out in graceful arcs until one marveled that mere water could support such a weight of steel. Sprouting from the quarterdeck of each battleship, a cage- or tripod-type mainmast soared up more than a hundred feet, matched by a foremast of identical height rising from the multidecked superstructures. By twos and threes, tapered gun barrels nearly sixty feet long emerged from wedge-shaped, flat-topped turrets. The sight was greeted by a chorus of "Jesus Christ!" and "God damn!"—awed and inchoate tribute to one of the most dramatic works of man.

Even after the *Neosho* made another hard right into Southeast Loch and turned her fantail to Ford Island, we kept feasting our eyes on Battleship Row, now seen from abeam.

"Which is the *California*?" I had to know. It was identified as the one at the head of the line, nearest the sea. The petty officer turned to go.

"Looks like we're headed for Merry Point Fuel Dock," he concluded. "Time to get my gear together."

Thanking him for the guided tour, I looked at my watch. It was shortly after 1200. By now, tugs were at our port bow and quarter, nudging us toward a dock that was nearly surrounded by large oil-storage tanks. Only the vaguest breeze stirred; the temperature and humidity were both around ninety, we guessed. If this was paradise, why were we sweating?

# CHAPTER 4

# Life in the USS California

"Mason, radioman third class, reporting aboard for duty, sir!"

I had just ascended some forty steps of a mahogany accommodation ladder with a seabag and hammock balanced on one shoulder. It was not easy to look military while depositing the heavy burden on the quarterdeck, executing a right face to salute the flag and an about face for the salute to the junior officer of the watch. I was glad I had mentally rehearsed these movements a dozen times.

The ensign who returned my salute was in crisp, starched whites. He was wearing a service .45 and looked as if he might know how to use it. "Very well. Fall in and present your orders."

Fisher, Stammerjohn, and I formed a rank alongside our possessions. Our papers were passed punctiliously from the duty petty officer to the JOOW to a seaman messenger, who started forward at a dead run.

The expanse of teakwood deck on which we were standing had been holystoned with such loving care that it appeared bleached. Everything was freshly painted in pale battleship gray, including the turrets and their projecting guns, the two aircraft catapults that housed the scout seaplanes, and the double-angled crane at the fantail. Behind me, the cage mast changed paint schemes halfway up, from haze gray to black and back to gray again at the battle top. All the brasswork on hatch covers, ventilators, and ready-ammunition lockers glittered in the afternoon sun. The sailors were dressed in clean white pants and T-shirts—apparently the uniform of the day here in the near tropics. They went about their duties with no loafing or skylarking and a minimum of conversation. Everything seemed so taut and shipshape that it was hard to believe the *California* and *Neosho* (with the exception of her radio shack) belonged to the same navy.

Eventually, a tall enlisted man in his early or middle thirties appeared. While I couldn't tell his rate, he looked regular navy, distant and bored. He surveyed us with a pained expression and actually sniffed in disdain. An angular six feet two or three, he had a bony face smudged with blotchy freckles. Large ears and close-cropped sandy hair showed under his white hat, which he wore square across his eyebrows.

"Follow me," he ordered, heading for a door at the break of the boat deck. Shouldering our baggage, we hurried down a long main-deck passageway and made a right oblique across an athwartships space that was obviously a crew's messing area, judging by the tables that were triced up to the overhead. Passing a double line of men at a soda fountain and ship's store, we went down a wide ladder to a second-deck berthing compartment. Still moving forward, we descended a second, narrower ladder and entered another crew's compartment fronting the curved bulkhead formed by number two turret barbette. A short jog to starboard and we turned aft, down a passageway devoted principally to a long ammunition conveyor belt and its related hoists.

"Where are we going and how the hell would you ever get out of here?" I asked myself. Just then, our leader turned hard aport through a watertight door and we entered what would be our home for the next fourteen months. It was about twelve feet wide and thirty-eight feet long but looked larger because everything was painted a cold and sanitary white. Rows of two-high bunks and lockers were fitted around the seven or eight supporting stanchions. The only contrast to the white bulkheads, overhead, lockers, mattress covers, and folded blankets on the bunks were the dull-red navy linoleum on the deck and a wooden table and two benches in the center of the compartment.

We had been told that the battleship navy was tough and disciplined, and the compartment reflected those qualities. It was well below the waterline and the air was heavy and still. From the day of the *California*'s commissioning, nineteen years before, no ray of healing sunlight had ever penetrated here. It smelled faintly of men who sweated freely and showered daily, of painted steel and fuel oil, and of gunpowder.

The leading petty officer—for that, obviously, was what he was, although he had not introduced himself or even asked our names—stopped by the table and looked around in some perplexity. He removed his white hat and scratched his head. Then he pointed casually toward some unoccupied bunks on the inboard side—all uppers, of course. "Take your pick," he said. We staked our claims by tossing our seabag ensembles atop the metal springs. Allocation of lockers followed in the same fashion.

"How about chow?" Stammerjohn boldly inquired.

Apparently, the thought of food had never crossed the petty officer's mind. He looked at his watch. It was nearly 1600, and we had not eaten since breakfast aboard the *Neosho*.

"Yeah," he said absently. "Well, follow me." Relieved of our gear and assured of a place to sleep, we pursued him with great expectations back to the main deck and the messing area.

Aerial portrait of the USS *California* at anchor in San Pedro Bay, April 1935. With modernization postponed because of the international situation, she had changed but little when the author reported aboard in October 1940, except for the addition of an CXAM radar antenna atop the superstructure. (Photo courtesy of the National Archives.)

We found ourselves at a table apparently assigned to transients and early watch standers. It was presided over by a deeply tanned sailor who looked just as rough as boatswain's mates were supposed to. I figured that was his rate because he was wearing a bosun's pipe on a cord around his neck, just as my high-school coach wore his whistle. Like our leading PO, he was regular navy through and through. He, too, was over six feet tall, with biceps and chest that strained against his T-shirt. And he shared something else with the freckled one: he despised third-class petty officers of the naval reserve. He fixed Stammerjohn and Fisher and me, in turn, with a flat, malevolent gaze, as bleak as a cop eyeing three burglary suspects.

"Reserves," he said. It was not a question. "Well, you're in the regular navy now." He turned to his plate and ignored us for the rest of the meal.

I felt very young and vulnerable, as if I had been transported back ten years in time to a typical Mason Sunday dinner with guests present, where children were expected to be seen and not heard. I kept my eyes on my plate, reached discreetly for the tureens of food provided by the seaman mess cook, making sure the fearsome bosun's mate had first choice. It wasn't that I was physically afraid of him, even though he was ten years older and outweighed me by fifty pounds and probably could have taken me easily in a fight. It was because he was navy and represented what I at that time desperately wanted to be. Beyond that, only a fool would get into trouble with a superior petty officer in his first hour on board a battleship. The whole atmosphere said, so loudly that words were not needed, "If you step out of line, sailor, punishment will be swift, it will be certain, and it will be severe."

Of all the meals I had in the navy, that one remains the most sharply etched in my memory. The hard benches of the mess table. The six impassive faces, silent except for an occasional request to pass the bread or butter. The buzz of background conversation from nearby, more convivial tables. The metal lattice-work grille across the closed soda fountain and ship's service store, to my left as I faced the stern. The heavy, starchy meal. The sultry air. The hostile boatswain's mate intimidating everyone with his glowering presence.

Yes, I had met the battleship navy.

While we awaited our summons to the radio shack and our watch, quarter, and station assignments, we had an opportunity to get settled in and explore the ship. Not knowing what to expect, it was all the more a revelation.

I have read that large navy ships—battleships, carriers, and cruisers— could be likened to small, tight, self-contained cities of citizen-sailors working in close cooperation for a common cause. That is not exactly what I found in the *California*, whose crew might more aptly be compared to a collection of fiercely independent Scottish clans, reluctantly gathered together in the face of a mutual enemy.

In one sense, though, the *California* was like a small city—but not one I had ever known, except in the history books. The ship resembled a walled, fortified medieval town with a rigid hierarchical structure of peasants, artisans, clergy, and nobility, all answerable to the *seignior*, or lord of the manor.

The analogy is only a little strained. In a navy ship, it was possible to rise from "serfdom" to "master" of one of the craft guilds (divisions) and even, for the very few, to reach "knighthood." Still, as in the medieval scheme,

Seabag inspection in dress whites in the *California*. Leading the inspecting party in this undated photo is a four-star admiral. (Photo courtesy of the San Diego Naval Training Center.)

every man had his rigidly defined place, his specific duties and responsibilities. Strict observance of his obligations was duly rewarded, any breach of them sternly punished. He also had his rights, as spelled out in the Articles for the Government of the Navy, although it was not considered wise to exercise them beyond prudent limits. Indeed, the articles were known among sailors as Rocks and Shoals, or Death and Greater Punishments.

And it could not have been otherwise. The feudal structure in the *California* was the result of centuries of hard-earned experience in battle. Capital ships in almost every navy in the world were organized along similar lines, whether the country they represented was a democracy, like Great Britain and the United States, or a dictatorship, like Germany and Japan in 1940 or Russia at any time. It is a curious fact that, to survive, democracies must defend themselves with military organizations that are every bit as autocratic as those of their foes.

As I roamed the ship, I began to learn about her hierarchies and her twenty-three petty fiefdoms, or divisions. Each fiefdom was run by a chief or first-class petty officer from his command post near the ubiquitous coffee pot. Territory was guarded so zealously that my explorations were sharply proscribed. The bridges, ship and flag, were off limits. So were the conning tower, the turrets, and the various fire-control range finders and directors. The quarterdeck and the entire officers' country were off limits, except for enlisted men on duty. So were the warrant officers' and chiefs' quarters. Even my own living compartment was forbidden during daytime working hours.

If I can't go up to the bridges and battle tops, I thought, I'll go below. I was curious to see the propulsion equipment that drove this monstrous warship through the water. Someone told me that the *California* had a turbo-electric drive—boilers actuating steam turbines that rotated giant electric motors, which in turn powered the four screws—but I needed to see this machinery for myself and get a firsthand explanation.

When I went down a third-deck midships hatch into the arcane, hot, noisy world of the black gang, I encountered a first-class at his coffee pot. He looked up coldly, his annoyance unconcealed.

"What you want, Mac?"

Rather timidly, I explained that I just wanted to look around. His initial reaction ripened into hostility. Apparently, this was something that was not done.

"What the hell for?"

I had just come aboard, I replied, and wanted to learn something about the ship. The machinist's mate looked at me in astonishment.

"What the hell is your rate, Mac?" he demanded when he found his voice.

"A radioman third," I said.

Like a biologist trying to classify some strange new specimen, he examined me closely. Finally, he came to a tentative conclusion.

"Reserve?"

I admitted that I was indeed a reserve on active duty. Dismissal was immediate.

"Naw," he said. "I can't let you people go wandering around down here." Suddenly, I had been multiplied into a horde of rude interlopers. "Go see your division officer."

After the morning muster, I approached one of the junior officers of the C-D (radio) fiefdom. White hat in hand, I gave my name, rate, and request. A puzzled look crossed his face.

The ensign was slim, handsome, and well-scrubbed, his cheeks glowing pink with health. I judged him to be no more than twenty-two, a reserve who was the product of an Eastern prep school and one of the Ivy League colleges.

"What is your name again?" He pronounced it *ah-gain*, as President Roosevelt did. I told him.

"Mason," he repeated, as if filing both it and its bearer's foolish request for future reference. "Well, Mason, I'm afraid that's out of the question. Simply out of the question."

"Why?" I wanted to ask. But I knew better. If an officer denied a request—even this officer, who probably had little more time in the navy than I—that settled the matter. I hadn't yet learned that a junior officer might be asking for trouble if he interfered in the affairs of the all-important guilds. There were very practical limits to noblesse oblige.

With most areas of interest both above decks and below barred to us, Fisher and I often found ourselves at the ship's gedunk stand, despite the quality of ice cream served there. A thin, gooey concoction that tasted of dried milk solids and strong vanilla extract, slapped into a conical paper cup, passed for ice cream. Since it was unfit for human consumption without disguise, we added a nickel to the ten-cent price and ordered it dripping with chocolate or strawberry syrup topping. Never having heard of the dietary hazards of sugar and being unconcerned about weight—it was impossible for me, in those halcyon days, to gain an ounce no matter what I ingested—we ate great quantities of these horrendous sundaes until, faced with the cruel economics of sixty dollars a month, we switched to soft drinks at a third the cost. Our Cokes we disguised, too, with cherry or lime flavoring: a squirt of syrup into the already sugary drink. Probably the sole nutritionally acceptable item was Dole Pineapple Juice, but as I have reported, we thought the price outrageous, considering the juice was processed with cheap labor not more than seven or eight miles away.

The old-timers derided these effete practices of the new navy. "Goddam gedunk sailors!" they sneered, turning to their joe pots. But the gedunk stand, to us younger ones, brought back memories of the neighborhood soda fountains that were teenage social centers in our home towns, and there were always fifteen or twenty of us in the service line.

We consumed the sundaes and soft drinks standing up in the adjoining mess spaces or topside, in the five-inch-gun casemates or on the boat deck. The *California* and *Tennessee* were the first American battleships in which the five-inch, fifty-one-caliber secondary batteries had been installed in higher, dryer casemates, rather than on the main deck. On the earlier dreadnoughts—the *Oklahoma, Pennsylvania,* and *New Mexico* classes— these guns had been expensively relocated during major ship refits, after it was discovered that they were inoperable even in moderate seas in their original locations.

The fourteen slim, long-barreled guns of the secondary battery, with their elaborate elevating and training devices, looked truly formidable. No one told us that they had been developed and mounted in battleships before World War I, were useless against low-flying aircraft, and ineffective against even their designated targets—destroyers and other torpedo-carrying vessels—without the kind of sophisticated fire-control that no battleship of that time possessed.

But if the *California* was outmoded technologically, the men who ran her were not. Soon we were told to report to main radio, also known as radio one, for our meeting with one of the most highly regarded of these consummate professionals, Chief Radioman Thomas J. Reeves.

Radio one was located deep in the armored citadel, just inboard of the torpedo bulkheads and outboard of central station on the port side of the first platform deck. Only one entrance was provided, which took one down a ladder from the third deck and through an armored hatch whose weighty cover was propped open except during Condition Zed. The underside of the cover was ominously smooth, lacking either dogs or an escape scuttle. This meant that there was no way out once the compartment was sealed for general quarters, unless the hatch was undogged from above. If no damage-control personnel survived to perform this duty, the radiomen would die ingloriously, suffocated or drowned in their steel coffin. Filing that lamentable observation away for future verification, I followed my fellow V-3s into main radio, the nerve center of communications for the ship and the Battle Force flag.

We were in a sizeable compartment chockablock with operating positions around three sides and in a back-to-back row down the middle. Before I could finish my inspection, we were directed to wait in the adjoining communication office, which was staffed by a couple of ratings at desk and typewriter and two strikers who were exchanging liberty stories near a huge electric coffee pot. One of them was dressed for messenger duty, in undress white jumper with neckerchief, duty belt, and leggings.

Presently, a measured tread was heard on the ladder just above us. The messenger grabbed for his message board; the other striker fled to the radio side.

"The chief," the first one muttered, busily flipping through onionskin copies of messages to be delivered to officers' country.

As I straightened to attention, I had my first look at the man who would dominate my life for the next fourteen months. Chief Reeves was as I have described him, except that he was wearing a short-sleeved white shirt, open at the throat, and white trousers—uniform of the day for CPOs. After a quick look around main radio, he turned his riverboat gambler's gaze on his three reserves.

"So you're my new men," he said in a voice that was both sharp and commanding. A person with a good ear for dialects might have caught a trace of a New England accent. "Well, which one is which?"

Naturally, Stammerjohn spoke up. He gave his name, rate, and amateur-radio call sign.

"Ham operator," Chief Reeves said, lighting up a fat cigar. He seemed mildly amused. "I suppose you're a real hotshot."

Modestly, Stammerjohn listed his qualifications. The chief eyed his gaunt frame. He did not seem impressed. "You?" he asked, turning to me.

I said that my name was Mason, that I didn't know a thing about radio except what I had picked up at San Diego, but that I was ready, willing, and able to learn.

Reeves looked at me more closely. He nodded his head once, in what could have been interpreted as approval, before shifting to Fisher, who followed my lead.

The chief puffed on his cigar. I noticed a large diamond ring in a gold setting on his left little finger. The messenger slipped away on his appointed rounds; the striker carried cups of coffee to the watch standers. I felt muggy air on my face from the overhead ventilator, and heard the high piping rhythm of half a dozen radio circuits.

"All right," Reeves said. "The three of you will start as backup on NPM Fox. Your watch schedules will be posted."

Stammerjohn looked crestfallen. I felt relieved and grateful, as Fisher no doubt did. The day of reckoning had been postponed, and I had a little time to acquire the proficiency expected of my rating.

"Yes, sir!" I said.

A half smile formed itself on the chief's heavy, downturned mouth. "Don't call me sir, Mason. I'm not an officer." His tone left no doubt he was devoutly pleased about that.

He turned away, still with a half smile, then paused. "We're setting up practice sessions for you in emergency radio," he said. The smile faded. "Be there." He looked at Stammerjohn. "All three of you."

"All right, chief," we chorused.

We had just joined the ship's company.

In the early days of fleet radio communication, when most officers understood little or nothing of wireless telegraphy, navy radio operators developed into an elite group, practicing their mysterious arts largely free of military duties and answerable only to the commanding officer. Since skippers often had no more idea of what radio was about than their subordinates did, the enlisted radiomen found themselves in an enviable position—too enviable not to be exploited. Exploitation, as always, led to reform, and long before 1940, radio gangs were integrated into ship's complements. But the radiomen in the battleship navy still lived apart, in some respects, from the crew and had special privileges that many shipmates envied and some resented.

While the other divisions in the *California* stood the traditional four-hour watches, the C-D Division day was divided into four watch periods: morning (breakfast to dinner); afternoon (dinner to supper); evening (supper to 0100 the following morning); and "mid" (0100 to breakfast). Thus, the day was split unequally among two relatively short watches and two punishing ones, made more punishing still when we operated on a four-section schedule of alternating watches on the duty day and no watches the following, liberty day. (Later, when we had a surplus of radiomen, this system was modified.) After standing a mid, by special dispensation, we were allowed to "sack out" until dinner time. Otherwise, the living compartments were off limits from the time of muster on stations around 0900 until dinner time or later.

Fisher drew the first watch on NPM backup. When I went to relieve him after early chow—no longer at the table of the fearsome bosun's mate but in a special section of the main deck reserved for the radio gang—he was dead ahead as I came down the ladder, seated at the inboard bank of receivers. To his left was the regular NPM watch stander, also a third class. Just behind him was the watch supervisor, whose broad, cluttered desk looked into the communication office through a sliding glass window in the bulkhead. Obviously, we V-3s were positioned where we could be kept under close surveillance.

"Not too bad," Fisher said, rather pleased with himself. "Copy everything you hear. Supervisor will be checking you." He handed me the headset, scooped up his cigarettes, safety matches, and empty coffee cup, and was gone.

I seated myself in front of a large receiver, about two and a half feet square, tuned to 26.1 kilocycles (kc). (I learned later that all shipboard radio gear was limited to a little under thirty inches in the critical dimensions of width and depth by the standard thirty-inch-square hatch size.)

Since Radio Pearl Harbor's 500-kilowatt (kw) transmitter was located only a few miles away, the signals were as crystal clear as if I were copying a code machine.

The Fox schedule was the primary means by which CinCUS maintained contact with his commanders and ships afloat. Messages were always in code, except for the few personal ones still allowed in 1940, and were transmitted blindly, one after another, on a 24-hour-a-day schedule. Since the ships did not acknowledge receipt of Fox messages, it was impossible for hostile direction-finders to get a fix on their location. When transmitting blind, however, the assumption had to be made that the messages were received, regardless of atmospheric conditions or equipment breakdowns. Therefore, each unit of the heading—the originator, the date-time group, the action and information addressees, and the priority designator—as well as the five-letter groups of encoded text was repeated. Every message was thus sent at least twice, important ones more often. Having a backup operator made equally good sense, as insurance against the rare occasion when the regular operator lost the signal or had a typewriter jam. And it provided valuable training for inexperienced radiomen.

When a message came in with the *California* or the Battle Force commander as an addressee, the operator reached for a message form with several color-coded carbon flimsies, and cranked it into his mill over the rolled-up log sheet. If it had a priority or urgent precedence, the watch supervisor would fill in the heading—usually by referring to the backup operator's log—and fire it off to the coding room by pneumatic tube.

This "black chamber" was located just off the comm office, but I never got a look inside. On the locked door was a Secret—No Admittance sign. The small compartment held the navy's top-secret electric cipher machine, which was operated like a typewriter by C-D Division officers. The plain-language text emerged in long strips that were pasted down on message forms, much like Western Union telegrams, retyped, and delivered by officer-messengers. No enlisted man (except Reeves, apparently) ever saw the deciphered texts.

Messages of lower-than-secret classification were usually sent in a strip code, the first and last groups identifying the particular coding strip to be used. These were passed across to the comm office, where they were decoded by the radioman or yeoman at the write-up desk, retyped, and routed to the officers or offices concerned.

In port, main radio was not the crowded, bustling place it became at sea. Most of the radio circuits were secured, and messages were sent by visual means or by guard mail. This was especially true on the long evening watches and midwatches. The international calling and distress frequency, 500 kc, which we guarded for the Battle Force, was put on a loudspeaker. The basic ship-to-shore frequency, 355 kc, was manned, as was 2,716 kc, the frequency CinCUS used to communicate with all navy ships in the Hawaiian area. The training circuits, which allowed apprentice radiomen from all the battleships to practice sending and receiving messages to each other under simulated operating conditions, had yet to be inaugurated.

The people who ran decoded messages around the ship were the seaman strikers training to qualify for the third-class radioman examination. They were the low men in the pecking order, ranking even behind the V-3s. Craighead had been absolutely right about that. On watch, strikers had two other overriding duties. First, they kept the coffee pot going so that the watch standers, unable to leave their typewriters without an authorized relief, could have steaming cups of joe, or mud, on demand. (I don't recall hearing the word *jamoke* as slang for coffee in a navy ship, although it appeared regularly in the columns of *Our Navy* magazine. No American sailor ever uses a two-syllable word when a graphic one-syllable expression is available.) And second, strikers had to awaken the midwatch, a task that brought them verbal and occasional physical abuse from petty officers who had had but two or three hours of sleep.

When strikers were not on watch, their duties were even more menial: they cleaned compartments, polished brightwork, and served two-week stretches as messcooks. Some of our strikers had come up from the deck divisions in the time-honored navy system of shipboard training for the specialized ratings. For them, their hard lot was an improvement. But others had been to the four-month regular navy radio school and were better trained and more competent than Fisher and I.

On the evening watch there were always radiomen, mostly seconds and firsts, who sat around drinking coffee and "shooting the breeze." Since NPM did not run continuously in late 1940, I overheard some of their conversations with the watch supervisor, relaxing at his command post by the pneumatic tubes and the "patch board" that enabled him to plug "hot" transmitters in radio two to any of the operating positions.

Their talk was of a navy that would, within a little more than a year, be lost forever. It was not a large, impersonal navy then. Any petty officer with two or three hitches was almost certain to have served in the same ship or at the same shore station as some of the others. Consequently, the circles of mutual friends and acquaintances were wide and interlocking ones.

It was a navy that enabled one to see the world, or at least that part of it where the United States had important commercial and, therefore, political interests: Guantanamo Bay, Cuba; San Juan, Puerto Rico; Balboa and Panama City, Panama; Manila and Cavite in the Philippines; Hong Kong; and Shanghai. In the States, there were the great duty assignments of the East and West coasts: Boston, New York, Washington, D.C.; Long Beach, San Francisco, Seattle.

One or two of the radiomen remembered the good-will cruise of the

*California* and the rest of the battle fleet to Samoa, New Zealand, and
Australia in 1925, and the full-dress presidential review of the fleet off
Chesapeake Capes, Virginia, in 1930 (the place where French Admiral De
Grasse defeated the British Fleet under Admiral Graves in 1781, paving
the way for American independence). Listening while NPM marked time
with its call letters and *dit-dit-dit-dahs*, how I envied these old-timers for
their travels and experiences and sophistication!

Being professionals, their conversation was by no means limited to
sights they had seen, ports and women they had known. Much of it was
shop talk, but shop talk of a piquantly navy kind. They spoke of the annual
fleet problems in the Caribbean and off Panama, including Fleet Problem
XIV of 1933, where the "Black" forces had launched a successful carrier-
based air raid on Pearl Harbor. They talked of the communication officers,
execs, skippers, and admirals they had known, often in a personal and
intimate way that belied the gulfs of rank and class that lay between officers
and petty officers in the regular navy. Respect was there, the acknowledgment of military protocol and what one described as the built-in loyalty of
an academy officer to his men, but it was tempered by the realization that
officers were mortal, subject to the same weaknesses of flesh and spirit that
petty officers themselves were.

Some of these weaknesses were detailed with relish, especially the
drinking habits of certain three- and four-stripers, even admirals, and the
sexual practices of a very few of the officers' wives. The exploits of fortunate
and daring enlisted men, who had risked twenty years at hard labor in
Portsmouth Naval Prison to enjoy liaisons with women who made no
distinction between bands of gold and eagles of silver, were discussed with
wonderment and approbation.

There were conversations, too, about the hot radio circuits that the
eager amateurs from the San Diego radio school could only fantasize about,
such as the high-speed links between NAA Washington and West Coast
shore stations NPG San Francisco, NPL San Diego, and NPS Puget
Sound. These circuits handled so much traffic that a speed key (a semiauto-
matic keying device that permitted, once mastered, much higher speeds
than could be managed with a hand key) was essential.

Afloat, the top circuit was 386 kc, the flag tactical frequency used by
CinCUS in the *Pennsylvania* to communicate with his force and type
commanders. Only the most skilled operators qualified for 386, especially
during a fleet problem, when coolness under the stress of simulated
combat was just as important as a high level of proficiency. A challenge of a
different sort was NAX Canal Zone, the only navy radio station that
handled commercial traffic exclusively. NAX operators relayed radio-
grams, broadcast weather and hydrographic reports, and handled SOS
emergencies from ships at sea on several different frequencies.

Then NPM would break in with another numbered message and I would have to leave their fascinating world for a ninety-group deferred-priority communication from Commander Base Force to several World War I–era supply ships and oilers, transmitted at eighteen groups per minute.

As was customary with battleships and cruisers, the compartment that housed the radio transmitters was well separated from main radio. If radio one were knocked out, transmitters could still be tuned and vital messages sent from radio two. To insure two-way communication, the receiver at the operator's position in the transmitter room was powered with its own AC generator.

Located just forward of number three barbette, on the third deck, the transmitter room was protected on either side by the outer shell and five torpedo bulkheads, which together formed five narrow, longitudinal compartments—the inboard and outboard ones void and the other three filled with fuel oil. The overhead protection was less imposing but still impressive: a light splinter deck (the ship's main deck) and an armored second deck six inches thick, composed of two strakes of seventy-pound steel alloy.

On my first visit to radio two, I had been turned away. Now that I was a watch stander, I hoped for a friendlier reception. By battleship standards, it was a very large compartment, nearly twice the size of main radio, and seemed even roomier because the ten or twelve transmitters were well spaced. Being on the fringe of officers' country, the bulkheads were pale green rather than the white of radio one; the decks were covered with the standard thick red linoleum.

Where main radio seemed charged with nervous energy, even in port, the atmosphere here was relaxed. Everyone was in the dungarees forbidden the rest of the radio gang except during Friday field days. One sailor was operating test equipment at a large work bench; another was tuning up a transmitter; two or three more were lounging around the supervisor's desk drinking coffee. Forward on the first platform, the equipment was subordinate to the men. Here, clearly, the men were subordinate to the equipment.

As I stood there, a stocky, red-haired radioman gave me a hard, challenging grin. I had seen him before in my berthing compartment but we had never talked. "By God," he said, "here's one of the prima donnas from main radio. You get lost finding your way to the gedunk stand?"

By now, I was growing accustomed to being hazed wherever I went. This fellow, at least, was good natured about it. I grinned back and explained my purpose.

He feigned surprise. "You mean you really wanta learn something about the equipment you operate? By God, you're the first one. You're not kidding, are you?"

Assured I was serious, he got up and stuck out a meaty, callous hand. "Red Goff," he said. He proceeded to take me on a comprehensive tour of radio two, a technical briefing punctuated with wisecracks, off-color jokes, intimate revelations about women he had known, and the firm opinions that one, most of the operators up forward couldn't tell a vacuum tube from a condenser; two, the *California's* officers, right up to the admiral, were the worst collection of seaweed politicians and misfits ever assembled in one ship; and three, Honolulu was the absolute armpit of the world.

Aside from schemes to get transferred off "this horseshit flag" and onto a destroyer, Goff's enthusiasm was reserved for his radio gear. I, too, was impressed. Each transmitter was of rack-and-panel construction, about thirty inches square and six feet high, in a black crackle finish. Two or more transmitters of several models—designated TAQ, TBA, and TBB—were provided so that ComBatFor could communicate with his forces across the entire frequency spectrum from about 200 kc to 30 mc. Each transmitter had its own DC motor-generator set, which supplied up to 5 kw of power to the filament heaters and plates. Overhead, a double-bus transmission system loaded the output of the transmitters to the ship's antennas. A patch panel, similar to the plug-in telephone switchboards found in hotels of that time, was used to turn over control of the tuned transmitters to main radio.

It was well worth a little hazing to make a friend of Radioman Third Class J. F. Goff, one of the toughest, coolest men I ever met, a member of that rare breed to whom fear is apparently a stranger. Like the late Audie Murphy, the most decorated soldier of World War II, Goff was born for combat, but again like Murphy, he was insubordinate to a high degree and unable to find common ground between a military service whose regimentation he disliked and a civilian world he detested on principle. While he was quick to take offense and a formidable adversary, as I shall relate, he and I never had a serious disagreement—perhaps because I always found his irreverent and sardonic humor so refreshing.

As I was leaving radio two that day, he said: "You'd better transfer back here, Mason. This is the safest place on the ship. Any bomb that gets us will have to come through officers' country, and our goddam exec has just issued a directive putting the wardroom off limits to the enemy!"

It was when the training sessions for the V-3s began that I found a leisure-time hangout. Emergency radio, or radio three, was inboard of our living compartment, on the other side of the ammunition passageway, and just off the machine shop. It was a small, oblong space, scarcely large enough to contain all the radio gear that would be needed—a transmitter with its motor-generator; an operating position with a receiver and related 110 VAC generator; and the last resort, a battery-operated, portable transmitter-receiver—if the *California* ever got into such desperate straits that radio one and radio two were both out of commission. Adding to the congestion was training equipment, such as the code machine, practice oscillators, and typewriters.

There was also an aluminum percolator, for it was here that the junior chiefs and senior firsts gathered. This battleship navy practice was well described by Captain Cary H. Hall, USN (retired), then an ensign and Sixth (Port) Division officer, in a letter to me dated 26 April 1980.

> Every division had a spot in the ship where there was coffee and the real business of the division—watch lists, leave lists, etc.—was done. There were enough corners in a battleship for this. . . . The trick for a junior officer was to be accepted in the "hidey-hole" as an equal, which meant being one-up on the foul-ups in the division, since the backbone characters would give pretty good indications on who was worthwhile and who should be shipped out on the next draft for lawn mowing at Mare Island.

The only difference between the deck division "hidey-holes" and radio three was that no C-D Division officers were in evidence in the latter. Anything that was decided in radio three had only to be approved by Chief Reeves to acquire the force of law. About the only time we saw our officers was when they were entering or leaving the coding room, at general quarters, and on the port quarterdeck at the Saturday morning captain's inspection. If we had a problem, we went to the chief.

It should be noted that all of the officers of the C-D Division were reserves. The only regular was Lieutenant H. E. Bernstein, communication officer in charge of C-D, C-L (signal gang), C-C (yeoman), and C-K (storekeepers and ship's service) divisions. Few academy men wanted anything to do with communications. Sitting in a stifling, closet-sized compartment at the keyboard of the cipher machine, aside from the sheer boredom of it, was not an accepted road to advancement, and was better left to Wall Street clerks, yachtsmen, professors of English literature, and the like. For the "admiral strikers," the action was on the bridge and other topside locations, not below on the first platform.

During the morning planning conferences of the senior petty officers, radio three was off limits. In the afternoons, Fisher, Stammerjohn, and I reported for code practice. These instruction periods were usually supervised by Schrader, the leading petty officer who had met us on the quarterdeck that first day. His approach was somewhat different from Craighead's.

"In this man's navy," Schrader said coldly, "a third-class radioman is expected to be able to copy thirty-five words a minute. And send twenty-five—without busting every other code group.

"If you V-3s have any ideas about making second class, you've got to be able to copy fast press, like you've seen Gus and Eck Sink and A. L. Smith do. You people will probably never get that good but, by God, you're going to be qualified thirds before I finish with you."

He turned the copy machine on and adjusted it for twenty-five wpm. "Copy!" he commanded.

"The hell I won't be that good, you bastard!" I promised him silently, and daily, as I struggled with speeds that always increased to remain just beyond me.

In the evenings and on weekends, emergency radio became my sanctuary. While most of the radio gang were occupied with liberty, the movies, or a concert by the ship's band on the quarterdeck, I spent pleasant hours there reading or writing letters home.

I would usually find a quiet, serious young third class at another typewriter. Finding common interests, we quickly became friends. M. G. Daley was the first regular I had met, Reeves and Goff excepted, who didn't seem to resent the fact that I was a reserve. He was one of the vast, masochistic tribe whose hobby was writing fiction. Since I had at least a dilettantish curiosity, we spent many evenings discussing the problems of writing for the New York pulp magazines.

For a while, I gave Daley a hand in the construction of plots for the westerns he was writing. Our collaboration didn't help him. He would send off a short story by first-class mail, and in scarcely more time than the round trip required, it would be returned with a form rejection slip. I was soon convinced that writing fiction was about the most difficult and frustrating of all activities, a wholly one-sided transaction in which the writer invested his time, effort, and expenses against astronomical odds of any financial return.

But under his diffident surface, the wispy-mustached Daley was a dauntless fellow. When I last saw him, some months later, he was still acquiring rejection slips—and was still determined to be a successful free-lance writer.

While we were still moored to interrupted Quay F-3 at the head of the battle line, I met another radioman third of the regular navy. Over the next fourteen months Melvin Grant Johnson was to make a profound contribution to my education. Our friendship is still a cherished memory.

But he certainly exercised no influence, except possibly a negative one, that first day in compartment B-511-L. Johnson was, in fact, a prisoner at large. "Sweating out" a summary court-martial, he was forced to muster several times a day at the master-at-arms shack near the ship's service store. The story he told me was an incredible one, related in his inimitable machine-gun style, words pouring out in bursts up to two or three times the speed of normal conversation. He was a man who lived his life as he talked, and when he violated navy regulations, which was often enough that he had already lost his second-class rating, it was always done with verve and style.

On 22 September, Johnson related, he went to Honolulu on liberty. The *California* had returned to Pearl Harbor from Bremerton less than a month before, and his heart was still in Seattle. Finding a willing companion among the bar girls on North Beretania, he got drunk. Scuttlebutt had it that the *Arizona* was leaving for the States the next day, and he resolved to join her. He reported aboard with a radioman he knew and was given shelter in one of the radio gang compartments. The *Arizona* did in fact get under way the next morning but, in the sober light of dawn, he couldn't figure out a way to sustain the charade without risking a general court martial. He reported to the Jimmy legs and was immediately placed under house arrest. As soon as the *Arizona* anchored in San Pedro, he was taken under guard to the *Platte* and returned forthwith to Pearl Harbor.

I listened with the fascination that the exploits of malefactors have for the law-abiding. The certainty of failure would have prevented me from even considering such a rash act. I asked him how he ever expected to get by with it.

The soon-to-be-familiar, swashbuckling Johnson smile flashed. "Hell, Mason," he said. "I was stoned. I wasn't thinking that far ahead. All I could think of was how much I hated this frigging Honolulu and how much I wanted to see Eleanor." A scowl chased the smile. "You know why we're out here? That S.O.B. Roosevelt hopes he can tempt the Japs into a stupid move. We knock them out, and then he can steam to the rescue of the Limeys and the Frogs."

Where I came from, it was gospel that the principal Roosevelt haters were the rich and a few moss back conservatives like my foster-parents. I was rather shocked to hear such criticism of my idol from a man who, obviously, was neither one.

"You're not going to tell the captain *that*, are you?"

"Hell, no. The old bastard probably agrees with me—so he'd throw the book at me," Johnson explained with impeccable logic. "What happened, see, is that I was given a Mickey and rolled. When I came to, I was so groggy I took the wrong liberty boat. Imagine my surprise when I woke up, heading east on the *Arizona!*"

It was, I had to admit, a good story. Captain Bemis must have thought so, too. Johnson got ten days in the brig and a twenty-dollar fine. By the disciplinary standards of 1940, especially considering Johnson had been absent from his ship for two weeks, that was a mere slap on the wrist. Shortly afterward, a fireman third class got a more typical summary-court punishment of twenty days' solitary confinement on bread and water (full ration every third day) and a thirty-dollar loss of pay. His offenses consisted of shirking duty and insolence to a petty officer.

After my first few weeks on board the *California*, it was apparent that, while I was in the battleship navy, I was not yet of it. I would have to prove myself before I truly belonged. Meeting that challenge would mean more than I could have imagined then. It would mean comradeship of a high order. It would mean membership for life in what I later described as a harsh and mystic brotherhood of arms. It would mean a quiet, secret pride and a deep, inner satisfaction I would never quite lose, no matter what happened.

# CHAPTER 5

# Lahaina Roads to Puget Sound

For more than two weeks, the *California* had been lying inertly at her berth near the Ford Island administration building. But at last the morning came when her port and starboard accommodation ladders were shipped, her boats hoisted in and nested one inside the other, and her boat booms rigged in and secured. Flags and pennants in rainbow colors were bent onto signal halyards and run up to the fore yardarms, announcing her intention to quit Pearl Harbor and return to her natural element, the deep ocean.

During the extensive preparations for unmooring and getting under way, I alternated between the forecastle and the boat deck, anxious to observe the means by which this great battleship cut her tenuous ties with the land and threaded the labyrinthine channel to freedom. I knew that this experience must be savored to the fullest now, before repetition dulled the sharp impressions of discovery.

Hours before reveille, the black gang had been lighting off boilers and cutting them in on the main steam line. At 0530, Condition Yoke was set and the special sea detail was manned. On the navigation bridge the officer of the deck was busy.

"Permission to jack over main engines, sir!"

"Permission granted," said the captain.

In this procedure, the turbines were slowly rotated by mechanical means, to insure even heat distribution and prevent warpage before they were used for propulsion.

"Permission to test main engines, sir!"

"Permission granted."

Now a navy yard pilot came aboard, and two tugs chuffed up, taking positions alongside the port bow and port quarter. The eight large manila mooring lines and the one steel hawser that tethered the starboard side to the quay were singled up, slacked off, and let go. On the interrupted quay, a line-handling party from one of the other battleships took them in, careful not to let the spotless bights dip into Pearl Harbor.

I felt the tremor of the deck beneath my feet and heard the faint whine of powerful electric motors. The dirty brown water under the ship's stern was whipped to a frenzy as the inboard engine was backed, forcing the stern clear of the after quay. Just as the bow was about to scrape the forward one, the outboard engine was backed and the bow stopped its swing. Simultaneously nuzzled by the tugs, in response to shrill whistle signals from the pilot on the bridge, the ship was slowly maneuvered clear of the mooring and was soon under way at one-third speed. The tugs continued to keep her aligned with the channel as she made the big turn to port around Hospital Point. With blue water ahead, the YTs cast off, the pilot scrambled down the sea ladder and was picked up by a trailing harbor boat, and the captain took the conn. We passed the harbor entrance buoys exactly one hour after getting under way and headed south for gunnery exercises off Lahaina Roads.

Early that evening, the call to battle stations was not subject to misinterpretation by the rawest apprentice seaman. First came the high, thin shrill of the bosun's pipe, quickly followed by the urgent *bong-bong-bong* of the general alarm and that most stirring of all bugle calls. In peace or war, I have never heard the rising arpeggio of general quarters without a quickening of the pulse and a slight hollow feeling in the pit of the stomach. As the last falling notes died away, "Now hear this! All hands man your battle stations!" blared out over the ship's public-address system. Fisher and I went to main radio on the double.

As we tumbled down the ladder, others were racing up two steps at a time: the evening watch, just relieved by the "first string," was hurrying to battle stations in radio two or radio three, the foretop and maintop, the bridge, or in the case of strikers and malcontents, the ammunition-handling rooms. Your GQ station, I soon learned, was an infallible clue to Chief Reeves's evaluation of your abilities and attitude.

Several new circuits had been set up before leaving Pearl Harbor, and additional tactical circuits were being activated for general quarters, so all the operating positions were filled. Even the communication office was full of people: a communication watch officer, his yeomen, a write-up desk radioman, talkers, messengers. As Fisher and I stood looking around, we were spotted by the chief, calmly taking a few last puffs on his cigar as he conferred with his watch supervisor.

"Into the comm office, you two," he said. "Stand by in case you're needed." Obviously, we weren't needed; it was the custom of the chief to assign new men to main radio, where he could observe them closely under stress.

Looking up, I saw the massive hatch cover, with its dead-smooth underside, being lowered into the coaming and dogged down with muffled metallic sounds. We were now sealed in, and there was no exit until a repair party let us out. I momentarily experienced the terror, akin to that some people have when flying, of abandoning control over my destiny and having my life placed in the hands of others.

The blowers were shut down and no air entered main radio. The exhalations of many bodies soon turned our limited supply of oxygen rancid with sweat and heavy with carbon dioxide. Standing squeezed under the ladder, I had nothing to do but sip on a cup of foul coffee and register the sounds and sights.

"One for the flag!"

"Striker! Cuppa joe!"

"Get this out to ComBatDiv 2 right away!"

"Striker! Pick up!"

"Priority for NAFT!" (NAFT was the *California's* radio-call sign.)

"Supervisor! I'm losing my signal!"

"Striker!"

The heavy air was penetrated by sounds: the *whoosh* of air as a leather message cylinder was inserted into a pneumatic tube and suctioned away to the coding room or the bridge; the muffled plop of incoming cylinder falling into its padded wire basket; the discordant howl of a transmitter being tuned on the air; the driving staccato of Morse code pouring from a dozen or more headsets at as many speeds, tones, and modulations.

Amid all the tumult and the shouting, the operators responded according to their various temperaments: some hunched tensely over their mills; others cursing and fiddling with tuning and volume controls; a very few as relaxed as only those can be who are truly competent and whose pride it is to make the difficult look easy.

Against the forward bulkhead, one of the firsts was copying WCX fast press so the officers would have their mimeographed and stapled news of the world at breakfast. Periodically, he would stop typing and take a long swig of coffee or light a cigarette as the machine-fed plain-language transmission raced on at well over fifty wpm; then he would catch up in a flurry of flashing keys and settle back again nonchalantly.

On the outboard side, the radioman second on the 386-kc command circuit was fluidly sending a tactical message, the light behind his key glowing red. Sweat was dripping from his forehead onto the message form—this operator was known for his enthusiastic consumption of whiskey in Honolulu—and he impatiently mopped his brow with a handkerchief without losing a *dit* or a *dah*.

A couple of positions down, in the corner, the operator on the 500-kc distress frequency had a blank log sheet in his typewriter, a face drained of color, and a partly filled bucket next to his feet. A recent transfer from shore duty, this woebegone third-class was still trying to find his sea legs. Next to him, the third class on the weather and hydrographic frequency was doggedly typing five-number groups in the international weather code—an excruciatingly boring task, as I was soon to learn. At his desk, the supervisor was instructing radio two to set up another transmitter, using a telegraph sounder fitted with an alternating-current buzzer that made a hoarse, rasping sound.

But nothing could compare to the rasping sounds that came from the omnipresent Chief Radioman Reeves when he detected any departure from procedure, any inefficiency, even an incriminating hesitation. The radioman who got "reamed out" took his punishment in embarrassed, red-faced silence. Here where he was king no one—and that included the CWO—challenged the chief. And so undeniable was his proficiency, so sure his mastery, that no one seemed to resent it. The overriding concern was to please him by achieving a "four-oh" performance. It was the ultimate compliment one paid to leadership and ability. Watching him in action at my first general quarters, I was awed. And I would remain so.

The next morning after chow, I hurried aft to watch the launching of our VOS planes, used for scouting and the observation and correction of main-battery fire. The *California* carried the usual battleship complement of three planes. The tail surfaces were painted white, the BatDiv 2 color code, to distinguish them from the planes of BatDiv 1 (red), BatDiv 3 (blue), and BatDiv 4 (black tail markings). As a force commander's flagship, the *California* was also entitled to a fourth plane, which was painted dark blue and therefore dubbed the Blue Goose. The floatplanes were launched from two gunpowder-operated catapults, one on the fantail and one atop number three turret. It seemed to me that being launched from a catapult would be more fun than riding the Giant Dipper at the Santa Cruz Boardwalk, with the bonuses of an airplane ride and radio gear to operate.

When I arrived, the two catapults had been swung out to starboard, each bearing a sleek, two-position, single-wing OS2U Kingfisher riding a launching car atop the structural-steel framework. The whole contraption did not extend the full ninety-seven-foot four-inch beam of the ship. The planes, I guessed, had no more than fifty-five feet in which to get airborne.

The VOS on the fantail catapult revved up its engine to full throttle. There was a muffled explosion. The plane shot down the track and, in a little more than a second, was free of the ship. After a short level flight it started climbing and banking. It was followed closely by the plane on the turret catapult. Within five minutes the other two planes had been hoisted onto the catapults by electric cranes and were fired off at a speed, I was told, of seventy mph.

Since each plane carried a pilot and an observer, usually an aviation radioman, I determined to check out the flight unit. I soon got acquainted with a tall, personable young third class named R. K. Harrison, who told me the planes carried a compact transmitter-receiver that could operate on both voice and code channels. Pretuned, plug-in modules allowed the observer to change frequencies quickly. Most communication was by voice, but when the planes were spotting main-battery fire, the corrections were sent back by hand key in Morse code. The observers also had to be able to fire a machine gun, and a chosen few got a chance at flight school in Pensacola, Florida. It began to sound better and better, even without the inducement of flight pay. Which was precisely why there were no vacancies in the V Division and a long waiting list.

Without an "angel," I had about as much chance of getting into battleship aviation as I had of going to Annapolis. I couldn't even manage to hitch a ride so I could experience the catapult thrill for myself. Everyone in the unit wanted to go up, and there were always officers in the wings finagling a joy ride. I would have felt much better if I could have foreseen that, within a very short time, the development of radar fire control and the rise of the fast-carrier task force would relegate the whole practice of catapulting planes off battleships to a minor historical footnote.

After spending most of the next two days below decks at watch standing, general quarters, and training sessions, I came onto the forecastle on the afternoon of 30 October to a reward that only a handful of people in 1940 had ever enjoyed: the view from Lahaina Roads.

The *California* and *Tennessee* were anchored just north of the storied village of Lahaina, its cottages and bungalows half-hidden beneath groves of palms, banyan, and other tropical trees. Behind Lahaina, whose sleepy demeanor belied its wicked reputation, the island of Maui rose tumultuously to a series of volcanic peaks nearly 6,000 feet high. Checkerboard squares of sugarcane fields marched bravely up the slopes, only to be turned back at the higher elevations. But even these heights were dwarfed by the mighty volcanic mass to the southeast, its dome shattered and hurled into the stratosphere by some titanic explosion of the remote past. Called Haleakala ("house of the sun"), it is the world's largest dormant volcano, with an elevation of 10,023 feet.

To the northwest lay a long narrow island, shaped like a roughly carved wooden shoe, its ankle end a jumble of volcanic spires.

"Molokai," I was told. "That's where the leper colony is." Hastily, I crossed the forecastle to the port side.

Across the roads was the humpbacked profile of Lanai, a study in browns and yellows, like the foothills of California's Coast Ranges in summer and fall. Directly astern, the vagrant breezes were swirling clouds of red dust over the barren island of Kahoolawe. I turned to look again at Maui and around the horizon. The jagged peaks rose up into fluffy clouds. The islands were many shades of green, deep lavender, tawny yellow, and burnt red. Over the roadstead, ruffled by the trade winds, spread meandering surfaces of azure, sapphire, and cobalt blue. I had never seen anything quite so lovely. This is where the creation should have taken place, I thought. Not in the barren Middle East.

I made plans to visit Lahaina. Man had corrupted Molokai. What had he done to his other corner of this Eden? But soon the announcement came that no liberty would be granted. No explanation was given, but rumor had it we were leaving at daybreak for gunnery training.

"You didn't miss a damn thing," I was told in the petty officers' head. "There's nothin' to do over there but get drunk and fight—and there's not enough Kanakas to go around." (*Kanaka* means a native Hawaiian or a South Sea islander. Among navy enlisted men, the word was always pronounced *kuh-'nack-kee*, and included Orientals as well.)

I consoled myself with the thought that there would be other chances to visit the place where two dominant traits of the American character—the lusty, boozing, brawling one and the hard-working, straightlaced, puritanical one—had come into such violent and irreconcilable confrontation a hundred years before. But it was not to be. Within two weeks the British launched a brilliantly successful night torpedo attack on battleships and cruisers of the Italian Fleet in the harbor of Taranto. Admiral James O. Richardson, the tall, tough, spectacled commander in chief of the U.S. Fleet, decided that the open Lahaina anchorage was especially vulnerable and sharply reduced the time the fleet spent there.

Since all my duty stations were on the third deck or below, I saw very little of the gunnery drills. I did get topside long enough to watch training practice with the fifty-caliber water-cooled machine guns located in the foretop and maintop and on both sides of the armored conning tower. The eight machine guns with the Texas-longhorn-style handlebars comprised our entire close-in antiaircraft defense. The targets were sleeves towed by two of our observation planes at slow speed and low altitude. As I watched the red tracers arc up at the fat fabric sleeves, it seemed to me that the gunners took a long time to score any hits. I had no idea what criteria were used to select machine gunners, but mine would have been very simple: only country boys who had grown up with guns in their hands would be considered. By the time a young man got into the navy, it was too late to teach him how to shoot. Reluctantly, I went below. When I emerged we were at anchor in Honolulu Harbor with the *Tennessee* and *Enterprise*.

I got the impression the admirals preferred any place but Pearl Harbor. Having observed that it took one hour and the help of a pilot and two yard tugs to clear that congested and oppressive base, I seconded the motion. From our anchorage, the Aloha Tower and the Koolaus were dead ahead, certainly an improvement over our customary vista of the navy yard's hammerhead crane, the tall, slender water tower at adjoining Hickam Field, and the oil tanks on Makalapa Hill behind the submarine base. From the wharves where the sacks of raw sugar were being loaded came the heavy sweet aroma of molasses.

The *Oklahoma* and *Portland* stood in and anchored. Merchantmen flying various flags stood in and out. An inter-island steamer was moored near the central terminal. Japanese sampans in bright colors ventured by for a curious look, drawing in turn the close and annoyed scrutiny of the OOD through his binoculars. The water was not the murky green-brown of Pearl Harbor but an inviting blue. One felt like plunging into its cool depths, as the diving boys did for coins flung by tourists from the decks of the Matson liners.

From the forecastle, Fisher and I reminisced about swimming parties on the American River, near the historic gold-mining village of Coloma. In those carefree civilian days of a few months ago, it had been easy to get a date. If only we knew a couple of girls we could take to Waikiki Beach, we decided, Honolulu might not be too bad.

For the next few days we conducted tactical exercises off the Hawaiian Islands with the other units of BatDiv 2, the *Tennessee* and *Oklahoma*, and made independent firing runs for our machine guns and the four three-inch fifty-caliber rifles of our antiaircraft battery. Before we returned to Honolulu Harbor, a piece of exciting scuttlebutt raced through the radio gang: we were going stateside! This was the straight dope! It came, by indirect means to be sure, from an impeccable source: Chief Reeves. John R. Mazeau, then a radioman third class in the flag allowance, described in a letter to me (25 August 1979) how our intelligence system worked.

As you know, they never told us when we were going back to the States, but Reeves was always told. Whenever we suspected it was time, we'd all sit up on the forecastle and wait. About two days before we were to leave, Reeves would come up the hatch from chief's quarters (located forward, on the main deck) carrying his best blues, with about a yard of gold on the sleeve, and take them to the tailor shop to be cleaned and pressed. He never said a word; but he let us know.

*Anchor's aweigh, underway!* If there are three words in the English language that sing with more poetry, evoke more nostalgia, promise more adventure, I have yet to hear them. The meaning of the words is precise, as nautical language usually is. But surrounding them is a pearl-like accumulation of rich associations. Evoked are memories of drama, history, perilous undertakings boldly hazarded; of faraway lands, strange tongues, exotic women, lives and destinies crossed; and always of home, which all men dream of even when they know they can never return. Anchor's aweigh, underway for Long Beach, California!

As I paced the forecastle, I felt singularly fortunate to be returning to the West Coast so soon after meeting the battleship navy. Perhaps the President would change his mind about basing the fleet at Pearl Harbor. Perhaps I would have a chance to show off my uniform on Main Street in Placerville and walk my foothills in their crisp autumn golds. Perhaps I would meet a young lady who would make me forget my old love. And perhaps, now that I had been to sea, I looked older than my nineteen years and the bartenders would not ask for my ID card.

As we cleared Honolulu Harbor, I could see that many other ships were under way. We changed course and steadied astern on the *Tennessee* and *Oklahoma* in a column formation. The eroded acute angles of Diamond Head loomed up on our port beam. Across the Kaiwi Channel, the purple highlands of Molokai came into view, steeped in menacing haze.

Presently the three battleships, now in column open order, were joined by the cruisers *Nashville*, *Portland*, and *Savannah*, the carrier *Enterprise*, and no fewer than ten destroyers to form Task Force Three. The three cruisers took station in the van, followed by a screen of four Battle Force destroyers—the *Benham*, *Dunlap*, *Ellet*, and *Fanning*. The *Enterprise* and her six escorts formed up far astern.

As she pitched and rolled arthritically in the freshening seas, the flattop reminded me of a huge but decrepit old man with an entourage of body-guards. If some seer had told me then that the *Enterprise* would steam to glory on one of the most brilliant combat records of any ship in the history of the navy, I would have given him the pitying smile one reserves for fools. From where I stood, it looked like she might have trouble reaching the West Coast.

I suddenly realized I was almost alone on the boat deck. Going below, I could hear the shouts of the gleeful seamen reverberating from the white-tiled enlisted men's head, far forward on the main deck. In my compartment, dress blues and peacoats were being broken out and examined with critical eye. The Lotharios were leering over names and telephone numbers in their "little black books." The "politicians" were devising brilliant schemes that would reward them with seventy-two-hour liberties. The sophisticated, planning excursions to Hollywood, discussed what they would find at Ciros, the Mocambo, Earl Carroll's, and the Florentine Gardens. The real salts, like the second class who handled 386 kc, were planning binges of epic proportions. And, in the tailor shop, the stacks of uniforms awaiting cleaning and pressing reached toward the overhead.

Over the next three days I would learn something about the ever-changing moods of the ocean misnamed the Pacific. From great rising masses of cumulonimbus the ship was bombarded by intermittent rain squalls. Almost hour by hour the ocean changed personalities, from smiling to sullen to grim to angry. Whitecaps formed; spray blew along the wind;

the seas heaped up and began to break in a fury of white water. Unlike a surf marching in to shore, there was no disciplined movement. Waves that seemed as tall as oak trees, moving with the speed of freight trains, collided with each other in rolling shock and flung spindrift. The ocean resembled an expanse of the Dakota Bad Lands heaving in continuous earthquake.

At first the *California* breasted the seas with a long, easy roll. When assailed from all quarters, she shrugged the waters aside like a grizzly bear shaking off a pack of hounds. As the sea redoubled its attack she gave a little ground, dipping her mighty prow into the troughs, letting the green water roll up her forecastle to break in vain against the solid steel of turret. Then her bows rose triumphantly, tons of ocean fell away, she showed her black boot topping in disdain, and smashed into the next trough in a hiss and roar of defiance.

I grew to love the *California* in those three days when she was tested and washed clean, as if purified by storm. Then I began to understand the feeling a sailor can have for his ship. The *California* was honest and she was virtuous and she was valiant. What more can a man ask of his mistress?

There is nothing to mark the approach to San Pedro Bay from the sea save the low, indistinct outlines of the Santa Barbara and Catalina groups of islands to the north and south, respectively, and the even more nondescript coastline, only the cliffs of the Palos Verdes Peninsula giving it character. Even the harbor is artificial: a dimple in the littoral at the mouths of the usually dry Los Angeles and San Gabriel rivers, protected from the sea by the long, interrupted line of two breakwaters.

We entered the main channel, passed between the breakwaters, and anchored in column with the *Tennessee* and *Oklahoma* on the San Pedro side of the bay. I saw many happy faces among the petty officers as they gazed shoreward. This was the Pacific Fleet's original Battleship Row; returning here was like a joyous homecoming, and every landmark evoked nostalgia.

Before the accommodation ladders could be rigged, a boat came alongside and a group of civilians awkwardly clambered up the sea ladder. They were a customs and agricultural inspection party, I was told. It was strange to watch middle-aged men in business suits invading the ship. Already, I was thinking in terms of "us" and "them." I resented their presence, with its tacit acknowledgment that they had the power to intrude upon our sovereignty, and was happy to see them go. With a grin of understanding, I reflected that I was becoming as hostile towards civilians as the *California* crew had been toward me when I reported aboard fresh from boot camp.

From the boat deck, I looked around the harbor. To the west, along one side of the main channel, San Pedro was a warren of squalid flophouses,

Excellent 1939 view of the port of Long Beach, looking toward the sea, with Santa Catalina Island in the distance. Two divisions of battleships and a number of cruisers and other navy vessels are anchored in the lee of the interrupted breakwater. Note the forest of drilling derricks in the already congested port; oil had been discovered there in 1936. (Photo courtesy of the Port of Long Beach.)

Famous Shanghai Red Cafe in San Pedro, where the menu featured chow mein (Shanghai Red's sentimental gesture to his days in the Asiatic Fleet), but where the principal attractions were strong drink, sailors' girls, and trouble. The entire block was razed some years ago as part of a redevelopment project. (Photo courtesy of the Naval Institute Collection.)

bordellos, bars, and cafes. One of its principal claims to fame was Shanghai Red's at Sixth and Beacon streets, known to all sailors as one of the toughest joints in the world.

To the north, past the moles that protected the channels and basins of the inner harbor, the port of Long Beach was a congestion of wharves, piers, railroad trackage, transit sheds, and oil-storage tanks. Oil had been discovered four years before, and drilling derricks had sprouted like weeds after a spring shower. (In 1940 the navy had just acquired one hundred acres of swampland on Terminal Island, at a token price of one dollar, for construction of a new navy base and yard; it would not be commissioned until September 1942.)

Just to the east, Long Beach was Des Moines with an ocean frontage. But my shipmates, with a few exceptions, were blind to the ugliness around them. This was what made all the duty in the islands, the incessant and demanding and sometimes dangerous drills and exercises, the demeaning and exploitative treatment ashore, worthwhile. This was the States! All who were entitled to liberty (and a few who were not) lined up on the port quarterdeck in immaculate dress blues and thought nothing of the long, choppy motor-launch trip to the Pico Avenue landing. They were escaping the regimentation and confinement of a navy battleship for fifteen glorious hours.

Not many of them would find what they were seeking. But they had freedom for fifteen hours, and freedom is hope. Enviously, I watched them go and could hardly wait for my turn the next afternoon.

Anchored within sight of the promised land, the ship took on a decidedly holiday atmosphere. Watch schedules were revamped to permit maximum time ashore. Liberty expired at 0715, not the 0100 in effect at Pearl. At every mess table, fresh milk and buttermilk were served in large metal pitchers. Nothing had ever tasted quite so good. For breakfast, we had fresh eggs with rich golden yolks and whites that stood their ground. And the gedunk stand, wonder of wonders, stocked some real ice cream.

The "liberty hounds" began returning with grandiose tales of adventures in Long Beach, Redondo Beach, Santa Monica, Los Angeles, and Hollywood. I felt a little lost as I listened to their stories in my berthing compartment. My new friend Daley had friends in Los Angeles; Johnson and Goff did not invite me to accompany them; Fisher and Stammerjohn had different watch schedules. On my first liberty, I went ashore alone.

The navy landing was located between Piers A and B at the foot of Pico. Housed in weathered frame buildings under a common roof were the shore-patrol headquarters, a first-aid and prophylactic station, separate waiting rooms for officers and enlisted men, and a civilian-run cafe that sold hamburgers, coffee, and beer. The fleet athletic field was close at hand. So were many pylon-shaped oil-well derricks, and the thick, sweet smell of crude oil permeated the air.

On the other side of the nearby Los Angeles River channel were the Long Beach amusement park, the Pike, and its associated Silver Spray Pier, accessible in those days before the freeway by means of a long, wooden pedestrian bridge. For those in a hurry, taxis were always waiting at the landing, but I had no specific place to go, so I walked.

The Pike was a permanent carnival for sailors. Its choice of rides ranged from the sedate Hippodrome Merry-Go-Round to a famous roller coaster called the Cyclone Racer. There were the usual games of chance and skill, burlesque side shows, palmistry booths, gift shops, bars, hamburger and chili stands, and concessions dispensing snow cones and saltwater taffy.

Facilities of a more substantial sort included the Majestic Ballroom, where ships' dances were held, a huge saltwater swimming pool, and the municipal auditorium, a Victorian pile that stood over the water in the lee of the horseshoe-shaped Rainbow Pier. Sailors took their girl friends there to jitterbug to the music of Artie Shaw or one of the other big-name bands.

As I passed a large, wooden bandstand, the daily session of the Spit and Argue Club was under way. Mostly retired sailors and shipyard workers, they gathered there to play checkers and cribbage, discuss New Deal politics and the international situation in loud, belligerent voices, make dire predictions about the future of such a soft, undisciplined country, and reminisce about the days when the navy was really tough. I would have liked to stay and listen but wasn't sure of my reception. The old gaffers became increasingly hostile as the day wore on and the liquid level in their whiskey and wine bottles diminished, I had been told.

An amusement park, I decided, could be a very lonesome place. Not

In this fine aerial photo, taken in the early 1930s, the navy's famous Pico Avenue landing occupies the frontage between Pier A and the just-completed Pier B (*right foreground*). Just to the left is the fleet athletic field. The wooden pedestrian bridge across the Los Angeles River (*center*) took liberty-bound sailors to Long Beach, whose skyline is visible (*right background*). (Photo courtesy of the Port of Long Beach.)

even the high-pitched feminine shrieks that accompanied the rumbling Cyclone Racer on its swift downward plunges could keep me there. I went up several flights of steep concrete steps and found myself on Ocean Boulevard.

It was much like Broadway in San Diego, minus the boots and most of the grime. The floors above the locker clubs, pawn shops, tattoo parlors, jewelry stores, and navy tailors (one, "Battleship" Max Cohn, was a regular advertiser in *Our Navy* magazine) served as cheap hotels and apartments.

Two busy cocktail lounges made it clear by their names, the Cruiser and the Saratoga, that they catered to sailors. Nearby were the Chatterbox (reported to be a favorite hangout of Chief Reeves), the Copper Penny, and the Skyroom, atop the Wilton Hotel. Despite the fact that most of the bars in downtown Long Beach, as in San Diego, were owned by retired chiefs and warrant officers, all were off limits to a nineteen-year-old. I was standing on a busy corner near the civic center trying to decide what to do when I felt a nervous tug at my sleeve.

"Young man," came a shrill, quavering voice. "Please escort me across the street."

I turned to find a woman of advanced years, wearing a bizarre costume of many colors, taking my arm in a firm grip. Her face was chalk pale and

Sailors are greatly outnumbered by civilians in this undated photo of the Long Beach Pike, probably taken in the 1950s. In the early 40s, this waterfront amusement park was awash with blue uniforms. (Photo courtesy of the Naval Institute Collection.)

deeply creased. She had applied scarlet lipstick across her withered mouth and some kind of purple shadow over her sunken eyes. This is how one of *Macbeth's* witches must have looked, I thought, and was immediately ashamed of myself. I had been taught to respect age, and it wouldn't hurt me a damn bit to help this tottering old lady across the street.

It seemed to take a long time. Sailors crossing from the other side of the intersection gave me faint smiles of commiseration, obviously thankful it was I who was playing boy scout, not they. I finally got her safely across by bodily lifting her up to the sidewalk.

"My, it's a pleasure to be escorted by a handsome young sailor," she trilled, relinquishing my arm reluctantly.

I turned away, but not soon enough. Our slow progress had been marked by another beldam, who immediately seized me.

"Such a polite young man!" she exclaimed. "I'm sure you won't mind seeing me safely across."

There was nothing to do but retrace my steps, even more slowly than before. This one let most of her 300 pounds rest on me. All the time she kept up a running conversation about the speed and recklessness of the drivers, and how it wasn't safe for an old lady to be out on the streets anymore.

After the last truly laborious step up, I made the mistake of lingering for politeness' sake and was again impressed into crossing duty. Each time I felt myself the object of more curious stares from more enlisted men at the busy corner.

The little and not-so-little old ladies were waiting for me now, each one thrilled to be assisted by a deferential sailor. It had become a sort of game for them. Even as I prayed that no one from the radio gang would see me, I was not surprised that I had been singled out. The very old and the very young and stray dogs and cats had always chosen me when they needed a friend. No matter how forbidding I tried to look, they were not deceived. They sensed I was soft hearted and could be depended upon to help.

One uniformed loiterer watched me at these rounds with increasing amusement. Finally he came over. "This your duty station, Mac?" he asked with a grin.

"Looks like it," I said. "How about a relief?"

To my surprise, he nodded agreement. As the next grandmother reached for me, he said, "Ma'am, my friend has to leave. I'll see you across." And he did, talking to her animatedly, not a whit embarrassed.

Thanking him fervently, I escaped into a nearby coffee shop. This was hardly the sort of liberty one bragged about in the after radiomen's compartment. I would have quickly forgotten it but for a phone call I placed from the fleet landing.

For a long time I had been wanting to meet my great-uncle, the

Reverend Lyman R. Bayard, my father's uncle on his maternal side. If it seems strange that I did not know him, it is because I was the product of a broken home and subsequent adoption and was seldom permitted to visit any of my blood relatives, including my mother and father.

Like all children who have been adopted, I had a great curiosity about, and a keen desire to better know, my real family. My Great-uncle Lyman was pastor of a Methodist-Episcopal church in southwest Los Angeles, only an hour or so away by one of the Pacific Electric's Big Red Cars. When he invited me to spend the following weekend at the parsonage, I gladly accepted.

It proved a most worthwhile visit, one that had more than a little significance for my trials by blood and steel to come. At my great-uncle's Spanish-style home, in a study that was a blaze of color from copies of religious paintings, he and I began a friendship that would endure for the rest of his long life.

In 1940 he was already nearing sixty—practically senile by the myopic standards of youth. But that is not how he looked to me. He was over six feet tall, with a fine erect figure and scarcely a line in his handsome, patrician face.

To my delight, he talked at length about my ancestors. The Bayards, he said, traced their ancestry back to Pierre Du Terrail, Seigneur de Bayard (1474–1524), a French military hero of such valor and virtue that he was called *le chevalier sans peur et sans reproche* ("the knight without fear and without reproach"). So revered was he that his enemies wept at his death in battle during the Italian Wars.

I returned to the ship with that great phrase echoing in my mind: *le chevalier sans peur et sans reproche*. It would become the standard against which I measured my officers—and against which I measured myself. When war came, most of them fell as far short of this *beau ideal* as I did. Although I never admitted it, fear was with me often, and I had a nodding acquaintance with reproach. But the memory of the Chevalier de Bayard, with whom I shared a bond of blood, helped me fulfill my obligation to my shipmates and my country.

While the 1,200 officers and men of the *California* were celebrating our return to the States in as many different ways, the business of the fleet went on. The passengers we had transported from Pearl Harbor left the ship, a few men were discharged, others were transferred. Several large parties of boots reported on board for duty.

In the late afternoon of 19 November, the entire BatDiv 3, comprising the *New Mexico*, *Idaho*, and *Mississippi*, stood in and anchored. Riding in the *New Mexico* (the temporary flagship while the *Pennsylvania* was in Puget Sound for overhaul) was CinCUS himself, Admiral Richardson.

The bluff, outspoken Richardson, I later learned, was vehemently opposed to basing the fleet at Pearl Harbor ("a God damn mousetrap," he is said to have called it) and made his views known to his commander in chief in unmistakable terms. He was a courageous officer and, as history has proved, a farsighted one, virtues which were to cost him his job. But for a week or so in late 1940, it looked as if the fleet had moved back to the West Coast. Present in San Pedro Bay were six battleships; the huge (and unlucky) carrier *Saratoga*; the cruisers *Nashville*, *Portland*, and *Savannah*; the oiler *Tippecanoe*; four destroyers; and various other craft. The *Enterprise* and her escort were nearby in San Diego Bay.

The merchants of Long Beach must have rubbed their hands with glee, thinking that the good old days had returned. In their hot pursuit of young women, the enlisted men of the *California* were certainly doing their best to encourage that notion. Inevitably, a few were absent over leave each morning or showed up as reluctant wards of the shore patrol. Just as inevitably, they were put on report and mustered at the daily captain's mast.

Punishments were meted out with no exceptions. Sometimes they were mild: extra duty or deprivation of liberties. As the length of time AOL increased, so did the penalties, in draconian proportion. For reporting on board two or three hours late, brig time was assured, often in solitary confinement on bread and water. If the AOL approached a day, a deck court-martial was awarded; if more than a day, a summary court.

Languishing in the brig with Long Beach a twenty-five-minute boat trip away was punishment exquisitely fitted to the crime of violating one's liberty privilege. The prisoners could not even see the city, for the brig was located on the third deck, just forward of the entrance to main radio. All they could see were steel bulkheads and, through the grating in the heavy door, the head of a unfriendly marine sentry.

According to the horror stories that came up from the brig—amply corroborated by Johnson and others I knew—the proper adjective for the marine guards was *sadistic*. They woke up the prisoners several times a night for the sole purpose of turning their mattresses over. They forced them to run in place or do push-ups to the point of collapse. They harassed them in a dozen other ways designed to abase, from using abusive language to withholding of reading and writing materials—and even meals from those not on bread and water. If a prisoner was provoked into retaliating, the best he could expect was a working over by the marines with fists and clubs. He might then find himself on report for "insolence to a brig sentry."

After seeing the place, and hearing such tales, I resolved never to know it from the inside. Getting into trouble afloat or ashore, when this sort of retribution was impossible to avoid, seemed to me the height of stupidity.

As we neared the end of our stay in port, one of the most envied men on the ship was a seaman first named G. M. Michael. His assignment was to drive the ship's service truck to Bremerton, Washington. One of the least envied was a marine private who had been given thirty days solitary confinement by a summary court-martial board. The same day "Lucky" Michael left the ship, the private was released from the brig and turned in to sick bay. "Diagnosis undetermined. Sentence of Summary Court Martial not completed," the log noted dispassionately.

It was common knowledge among the crew that a prisoner had to be more dead than alive to be released from the brig prematurely. It was also known that the function of sick bay was to get such men well enough so they could be put back in "the slammer." The private's immediate future was not promising.

On our last day in port, a man under a different kind of sentence reported on board for duty. He was Ensign Herbert Charpoit Jones, USNR, destined to win the Medal of Honor one year and one week later. Posthumously.

We weighed anchor and got under way for Puget Sound on the morning of 2 December in company with our sister, the *Tennessee,* flying her two-star flag and the formation guide. As soon as we passed the breakwater and gained sea room to the west, we changed course to pass between the mainland and the Channel Islands.

All that day, and the next, I spent every free minute topside. Except for a brief period in the *Neosho,* I had never seen my native state from the sea; it was an opportunity not to be missed. On the first day, I was rather disappointed. To starboard, ten miles away, were sandy dunes, crumbling bluffs, and low coastal mountains. Ahead, a flashing navigation beacon marked the coast-guard lighthouse on tiny, steep-walled Anacapa Island. All in a row, marching to sea, were the tops of drowned mountains, islands named Santa Cruz, Santa Rosa, and San Miguel, indistinct in the ever-present southern California haze. This being a busy traffic lane, we passed many merchantmen and tankers on opposite courses.

The next morning we sighted the Farallon Islands, rocky outriders for San Francisco Bay. Bleak and windswept, they were fit abode only for seagulls and cormorants. I wondered idly how many shipwrecked mariners had perished there. We passed Point Reyes, guarding the bay where Sir Francis Drake had sheltered the *Golden Hind* in 1579. Then we met the fog, the same fog that had likely shrouded the Golden Gate so long before and prevented Sir Francis from laying claim to the magnificent bay that opened up behind it. How the great English sea wolf would have loved snatching that honor from the detested Spanish, who happened along in the packet *San Carlos* on a better day two centuries later!

This perpetual northern California fog, formed where warm ocean air drifts eastward across the cold California Current, ambushed us in a thick, damp, opaque mass that stretched full across the northern horizon. No sooner had the *Tennessee* vanished into its soundless depths than it pounced upon and enveloped the *California*, which slowed and edged in closer to the shore. The ship's fog siren began sounding its long-drawn, mournful, pulsating note. Like an echo, the *Tennessee*'s siren sound returned to us, seeming to come from all points of the compass.

For the rest of that day and the following one, the fog played its taunting game with us, lifting as if to let us escape, waylaying us again as soon as vigilance was relaxed. The *Tennessee* streamed a position buoy astern, but long before morning we had lost both buoy and ship. Late that afternoon, off Cape Blanco, one of the rain squalls for which the lonely Oregon coast is justly renowned came beating in from the northwest and cleared the fog long enough for us to sight our sister ship fifteen miles ahead. But she soon passed from view and we again found ourselves steaming independently. One of the radiomen assigned to the bridge tactical circuit reported that there were a lot of nervous officers on the navigation bridge.

"God damn!" he said in disgust. "Here we've got the new CXAM radar that can see right through fog and the *Tennessee* has got nothin' but a crappy old direction finder, and she's runnin' away from us.

"The old man doesn't trust any newfangled electronic gadgets. All he can think about is goin' aground and facin' a court of inquiry. Them Battleship Forty-three sailors are laughin' at us all the way to Bremerton!"

It was not until the morning of 5 December that the fog released us from its smothering grip. I had thought, at first, that a fog horn made a rather nostalgic sound, akin to the steam whistle of a railroad engine passing in the night, but I had long since abandoned this quaint notion and prayed I would never hear that deep and plaintive signal again.

After we passed well clear of Destruction Island and the other hazards to shipping along the rugged Olympic Peninsula, we made a long sweeping turn around Cape Flattery and entered the Strait of Juan de Fuca. When I was rousted from my bunk the next morning, no movement of the ship was perceptible. I headed topside with a cup of coffee.

We were anchored near a weathered scar that rose up from the waters of Puget Sound. Its name was appropriate: Blakely Rock. All around us were low, heavily wooded islands and peninsulas, probed by the exploring fingers of a host of coves and inlets. We got under way, stopped to pick up a pilot, and by 1000 were moored port side to one of several broad-beamed piers that jutted out into Sinclair Inlet.

I could see many long red-brick shops, some with tall chimneys, four dry docks of various sizes, and an impressive two-story brick residence crowning a hill to my left—the quarters of the commandant, I guessed. The city of Bremerton, cramped on its small peninsula, crowded up against the navy yard fences and gates. Just to my right, a single-ended wooden ferry was backing into a creaky slip. Sprouting from the pier and extending its hammerhead shape across the ship was the biggest crane I had ever seen. From somewhere close ahead, the faint booms of a saluting gun rose over the clatter and roar of machines and machinery. It was obvious that Puget Sound, unlike Pearl Harbor, was a real navy yard. And from all I had heard, across the sound, fifteen miles to the east, was a real city—not a heap of carelessly flung jackstraws like Honolulu.

It seemed to me that I had come a long way from San Diego and Pearl Harbor and Lahaina Roads and Long Beach. What adventures awaited me here in the Pacific Northwest?

# CHAPTER 6

# Christmas in Bremerton: High and Dry at the Navy Yard

"Here come the goddam yardbirds!"

A pair of heavy wooden brows (portable gangplanks) had just been put across from the pier to the port side of the ship. A large group of workers in hard hats and overalls, who had been loafing at pier side, now began straggling on board.

"What are they going to do?" I asked the speaker, a rail-thin boatswain's mate with a hard, sunburned face. I was at my customary lookout station on the boat deck.

"Goddam little," was the reply. "Mainly, foul up this ship so it'll take three months to get things squared away." He walked away in disgust, muttering to himself.

Some of the yardbirds were dragging heavy equipment: welding gear, oxyacetylene cutting torches, and tanks. Others were carrying drills, hammers, and tool boxes. Obviously, they were planning to transform our home by cutting, burning, and mutilation. The crew members eyed them sullenly and gave them gangway with reluctance.

Most of the shipyard workers were older men, their gaunt, wrinkled faces bearing witness to humiliations endured in the fierce economic struggles of the "boom and bust" 1920s and 30s. Knowing all too well what the Depression had done to the wage-earning class, I should have felt for them the comradeship of the dispossessed and hailed their good fortune in finding well-paying government jobs. But my loyalty to service and ship had already superseded my identification with a class to which I belonged only by adoption. I found myself eyeing the yardbirds just as sullenly as the most proprietary boatswain's or gunner's mate.

Not all of the yardbirds were in their thirties or forties, and for those peers near our age we reserved our deepest dislike. If they were any kind of men, our reasoning went, they would be in the service. Naturally, they didn't meet the navy's high standards, but the army would be happy to have them—or at least those who didn't have moron-level IQs, heart murmurs, or flat feet, and

who didn't have insanity, epilepsy, or homosexuality in their family backgrounds. By virtue of navy yard employment, they had achieved the status of
4-Fs and would be safe at home, making far more money than they deserved,
in any future hostilities.

Their hair was too long, their clothes were filthy, and most of them had
misplaced their razors. And we were sure that, beyond this, they were devout
cowards. They would seldom meet our provocative stares and ignored the
insults we delivered when tripping over their power cables, which snaked
everywhere, or shielding our eyes from the brilliant blue-white light of their
welding arcs, or dodging showers of molten metal from their cutting torches.

While the crew seethed over this assault on their home, preparations were
made to subject her to even greater indignities. On 11 December, two yard
tugs came alongside and secured to our starboard bow and quarter. A much
larger oceangoing tug edged herself between Pier 6 and our port quarter. All
lines were singled up and the service connections to the dock were broken. A
second large tug snubbed herself to our starboard quarter. Our lines were
taken in and, with the tugs providing the motive power and the steering, we
were backed clear of the pier and then pushed forward between Piers 4 and 5.

Presently we found ourselves approaching the low profile of a dry dock
filled with dirty-gray, oil-streaked water. While the principle of a graving dock
is simplicity itself—a monstrous, leak-proof hole in the ground, fitted with a
floating or sliding caisson and pumps that can rapidly drain it of water—
squeezing a battleship into one called for a high degree of skill and cooperative
effort.

Dry Dock Number Two, built in 1913 and still in use at this writing, is 867
feet long and 130 feet wide, coping to coping. The granite blocks of which it is
constructed descend in step fashion to the floor of the dock, 38 feet down.
Since the California was 624 feet 6 inches in overall length, with a beam of 97
feet 4 inches and a mean draft of 30 feet 7 inches, there was little maneuvering
room. Nor was there the slightest margin for error, considering that the ship's
keel had to come to rest exactly along the centerline of large wooden keel
blocks secured to the bottom of the dock.

After the tugs had butted the California's bow into the open maw of the
dock, they cast off and the dockmaster took over, assisted by muscular
riggers and a ship's docking party. With much manhandling of eight-inch
manila hawsers, run from bitts on the ship to dock bollards, in response to
shouted orders and hand signals from the dockmaster, the California was
hauled in and centered in her granite and concrete sheath. The hollow
caisson was closed against the waters of Sinclair Inlet, the pumps were
started up and, in a remarkably short time, the ship was resting securely (I
hoped) on the keel blocks in the rapidly draining dock. Additional stability
was provided by other wooden blocks that supported the ship's docking
keel.

Stern view of the USS *California* in dry dock at Hunter's Point, San Francisco. The absence of a seaplane crane and catapults date this photo to the early 1920s. (Photo courtesy of the Los Angeles Maritime Museum and Port of Los Angeles.)

Out of her element, the *California* was like some great, barnacle-encrusted gray whale stranded on a hostile beach. With her engines secured and her boilers down, life-support systems at dockside provided all her sustenance: fresh water and salt water, steam, electricity, compressed air, and telephone service. Lacking these, she was nothing but a mass of cold steel, useless gunpowder, and inoperative machinery. From a puissant man-of-war that, afloat, could have reduced the skyline of Seattle to rubble in a matter of hours, the *California* had become as helpless as Prometheus chained to his rock. Even the huge, slab-sided rudder that guided her looked pedestrian and awkward now, and the four bronze screws that propelled her pitifully inadequate to their task. I turned away from my harsh scrutiny of my once-mighty mistress with a kind of shame for these thoughts and went below, where the illusion of beauty and power could still be maintained.

But it could not be maintained for long. The yardbirds returned and renewed their assault. The sickly sweet smell of burning metal filled the compartments. Eerie lights and shadows danced where the helmeted welders fused new equipment to decks and bulkheads. Wherever air circulation was poor, acrid smoke gathered to form eye-burning smog.

Adding to the confusion and the din, working parties of seamen had been put over the side to clean the marine crustaceans from the ship's bottom prior to repainting it with anticorrosive and antifouling paints. The dying barnacles stank in the sunlight. The high-pitched *rat-a-tat-tat* of pneumatic chipping hammers reverberated throughout the ship, as if all the world's woodpeckers were concentrating their attack on one giant hardwood tree.

Despite the smoke and fumes, the blinding lights, the falling rain of white-hot metal, and the mind-boggling racket, the shipyard workers seemed to spend most of their time bumming coffee from the crew or hiding out in remote compartments. Occasionally we would find one stretched out on the deck, sound asleep. One sailor voiced the popular sentiment.

"Goddam yardbirds," he snarled. "They think they're still on WPA. Why doesn't the navy make 'em shape up and work for their frigging money?"

When the quitting whistle blew at 1620, our unwelcome visitors did indeed shape up—for a speedy departure. As soon as we reclaimed the ship, nearly everyone who was entitled to liberty went ashore. I soon discovered why.

Without her customary insulating blanket of water, the *California* turned almost as cold inside as the Washington winter nights were outside. Her bulkheads and decks were ice-cold; the tiny trickle of warm air from her ventilators rose and hung somewhere against the overheads. In our bunks, we shivered and cursed under two GI blankets. I was sleeping in a compartment cold enough, it seemed, to preserve a side of beef. A few radiomen, pampered products of temperate climates, pulled on their wool turtleneck jerseys or huddled under the extra warmth of their peacoats. Emulating the senior petty officers, I endured the nights with what I hoped was regular navy stoicism.

In those days in Bremerton there was a huge, red-brick barn of a place on Farragut Avenue where an enlisted man could watch a free movie, dance until midnight to a ship's band, drink cold beer at ten cents a bottle or fifty cents for a large pitcher, and even meet a girl—all without leaving the navy yard.

Bloch Center at Pearl Harbor was new and had better facilities, but the Craven Recreation Center was more relaxed and spontaneous and, consequently, a lot more fun. Of course, the abundance of young ladies—obligingly brought in by the navy in busload lots three or four times a week for the dances held by the various ships—had something to do with that. Even the beer was much better than the green, uncured stuff served in cans at Bloch, which had given me a headache and a sour stomach on my one visit there.

At the time, I had the impression that Craven Center was quite old. It wasn't—it just looked that way. It was dedicated in May 1937 by Admiral T. T. "Tireless Tom" Craven, a friend of the enlisted man during his forty-five year navy career. As navy yard commandant, Admiral Craven was instrumental in having an abandoned brick hotel, built for navy yard employees in 1919, transferred from Treasury Department to navy jurisdiction and rebuilt as a recreation center. Since most of the materials were commandeered in the middle of a depression (with perhaps a "midnight requisition" or two), the building and its furnishings had a certain antique charm.

My introduction to Craven Center came in the line of duty, while we were still in cold storage in Dry Dock Number Two. I was one of seven or eight junior petty officers detailed from the ship for movie patrol at the center. The Annapolis ensign in charge posted me at a side entrance with specific instructions.

"No one is to be admitted here, sailor, unless he has a pass. Some officers will be showing up in civvies. That's okay—but they must have a pass."

I could see problems ahead. "Even a captain, sir?"

"Even an admiral. No pass, no admittance. Understand?"

"Aye, aye, sir!"

Attired in undress blues with a guard belt and leggings, my sole means of enforcement a regulation night stick, I took my stand at the door. For an hour or so, all went well. The officers were in uniform and they readily showed their passes. I saluted each one smartly and let him enter.

I was beginning to enjoy my limited and temporary authority—the first I had been given—when a man approached at a quick-time pace. He was dressed in a well-cut gray suit of heavy tweed, topped by a sporty fedora with a red feather in the brim. I knew at one glance that he was a navy officer, and probably a high-ranking one at that. No civilian of any station would walk that way, nor was likely to have that reddened, weathered complexion and that indefinable air of authority. To me, even his large, jutting ears bespoke command.

As he came near, I straightened to attention. Nodding, he started to brush past, reaching for the door knob.

"Pardon me, sir," I said diffidently. "May I see your pass?" I added another "sir" for good measure.

The man stopped, left half-faced, and pinned me to the wall with fiery blue eyes. "I'm Admiral Jones," he snapped.*

"Yes, sir, admiral," I said. I sincerely hoped I was doing the right thing. "But I have strict orders not to admit anyone unless he has a pass."

"I left it in my uniform," the admiral rumbled, sheering out slightly. "Who gave you that order, sailor?"

"Ensign Striker of the *California*, sir," I said, discovering one of the many advantages of being an insignificant link in the chain of command. "He's in charge of the movie patrol."

The admiral's face and ears got a little redder. "Go get Ensign Striker and bring him here," he commanded.

Another dilemma arose. I was not supposed to leave my post. But it seemed the lesser of two evils. And I would have sworn by all that was holy that Admiral Jones was exactly who he said he was.

"Aye, aye, sir!" I closed the door discreetly behind me and hurried away to find Ensign Striker. Fortunately, he was lounging in a doorway, watching the movie. When I told him about my blockade, he outdistanced me in the race back to the entrance.

The admiral was pacing back and forth, like Nelson on his quarterdeck at Trafalgar. "Ensign Striker!" he roared. "Did you instruct this man to stop anyone who didn't have a pass?"

Stuttering, the ensign acknowledged that he had. But the chain of command was his salvation, too. Those were his explicit orders from Lieutenant Commander Buckinghardt, he started to explain.

"Ensign Striker," the admiral interrupted. A grin was breaking like dawn over his red Celtic face. "Your petty officer did exactly what he was ordered to do. Very commendable, ensign; he was right on the ball." With a wink at me, he strode past us and down the hallway toward the movie at quick time.

---

*I have given the officers involved in this incident appropriate *noms de guerre*.

Ensign Striker struggled to regain his composure. Straightening up, he tucked in his chin, cleared his throat, adjusted his tie.

"Good work, Sparks," he said dryly. I didn't dare remind him that Admiral Jones never did show his pass.

Later, I heard that Jones was a carrier admiral, part of the "inferior" brown-shoe navy. But I remembered how he had winked at me, sharing his disdain for petty regulations, and was impressed. Maybe carriers weren't so bad, after all. In contrast with Jones, our own admiral, Charles P. Snyder, seemed as remote as Zeus Thunderer on Mount Olympus.

A few days later, Daley, the fledgling short-story writer of the radio gang, invited me to meet him at a Craven Center dance. He was bringing his new girl friend over from Seattle and wanted me to join them. I decided to check out the facility I had been so zealously guarding.

The dance was in progress at a high-decibel level when I entered the auditorium. At the far end, one of the navy's excellent bands was belting out what would later become classics of the swing era: Glenn Miller's "In The Mood" and "Sunrise Serenade," Benny Goodman's "Stompin' at the Savoy," and Count Basie's "One O'Clock Jump."

On a dance floor roughly the size of two basketball courts, hundreds of sailors and their partners were engaged in many imaginative versions of the jitterbug, some exposing a good deal of hose, bare thigh, and lingerie. The Formica-topped tables around three sides of the floor were elbow-to-elbow with onlookers drinking beer. No IDs were checked, the navy logically assuming that if you were old enough to serve, you were old enough to drink.

As I made my way down one side of the floor, dodging gyrating couples, a white hat waving aloft marked the position of Daley's table. I had expected to find him with a typical sailor's girl, a little coarse and not too pretty. But this was no First or Second Avenue pick-up. His girl friend had hair that came away from her high forehead in soft, dark-brown waves. Only a nose that was slightly too long kept her from being beautiful; the rest of her features were refined and delicate. Her voice was low and chimed like sleigh bells. Even in this noisy and crowded hall, she seemed as poised as a commanding officer's wife presiding over the silver tea service.

Mary Jane was a secretary for a Seattle business firm, having recently moved from Kansas City. She did not talk much about herself, but I got the impression it was an unhappy love affair that had brought her west, so far from family and friends. In a city full of eager sailors, the unpretentious Daley was fortunate to have met her, I quickly decided. But then, she was equally fortunate to have met such a gentlemanly one as my friend.

Occasionally, M.G. and she would dance to "Elmer's Tune" or "Green Eyes" as I stood in line at the ship's service canteen in the adjoining lobby for another pitcher of beer. But mostly, he and I took turns telling her about our backgrounds, our lives on board ship, and our ambitious if vague plans for the future. She gave each of us, alternately, her undivided and flattering attention.

Engrossed in this stimulating colloquy, I was scarcely aware of the twirling, jiving mob or of the passage of time. When the lights went on and the band swung into "Anchor's Aweigh," Daley and I had to check our watches to confirm that it was indeed nearing midnight.

Mary Jane said good-by with a radiant smile and a warm pressure of her hand in mine. I watched M.G. escort her toward the ferry terminal with envy. I wondered if he would spend the night with her at her Seattle hotel. Many enlisted men gave their shipmates vivid—and perhaps totally fictitious—accounts of amorous conquest, but that was not Daley's way. He never told me, and I never asked.

As I walked along the black, dimly lighted asphalt toward Dry Dock Number Two in the damp, cold night, I felt lonely. Mary Jane was the first woman I had met since volunteering for active duty that had impressed me. Already, I was mildly infatuated with her, and I had to remind myself sternly that she was Daley's girl, and therefore off limits.

Some sailors lived by the code that all's fair in love and war. Those sailors had few friends. If there was anything more important to a man-o-war's man than the respect of his shipmates, I couldn't imagine what it might be. I wanted to keep Daley as a friend—and so I kept my thoughts about Mary Jane to myself. It was a decision I may have regretted on occasion, but would not have changed.

After a week in dry dock, the *California*'s bottom had been repainted and the repairs to sea valves, propellors, and shafting completed and inspected by the ship's chief engineer. On 18 December, the dry dock was flooded, a yard pilot came aboard, and four tugs stood by. The dockmaster and his riggers labored mightily at the many lines. Once floated out of the dock, the powerless ship was turned over to the tugs for the return trip to Berth 6-D.

That the yardbirds were waiting for us was no surprise. Unexpected, though, was the sudden transfer of our skipper, Captain Harold M. Bemis, to the naval hospital for treatment of some undisclosed ailment. We were even more surprised when the Saturday plan of the day called for the traditional personnel inspection. It was conducted by Commander Robert B. Carney, our executive officer, who became acting commanding officer in Bemis's absence. The log judiciously referred to this event as "quarters for muster and Commanding Officer's Inspection" rather than "Captain's Inspection," the customary term.

Among the old hands there were knowing looks and low conversations. They said it was proof Bemis wouldn't be back. No exec would conduct his own inspection if there was any chance his former superior would return. It would be considered a direct challenge to his authority.

Sure enough, when Bemis returned a few days later, it was only long enough to be relieved of his duties. He was ordered to Cavite in the Philippine Islands as commandant of the Sixteenth Naval District.

I considered this a promotion until I was given the word by one of the senior firsts in the radio gang. The Philippines, he said, was regarded by many as a place of exile, where the Bureau of Navigation shipped overage or infirm officers to serve out their time until retirement.

So isolated was Captain Bemis from his crew (as were most prewar battleship commanders) that I was neither glad nor sorry to see him go. The other third-class radiomen and I rated him as a typical "spit-and-polish" battlewagon skipper. We were very soon to find out what spit and polish really meant and look back on the Bemis regime with a certain nostalgia. But at the time our main concern was that the "all-nav" exec with the hard Irish face be quickly relieved as acting CO. We didn't know when we were well off.

Whether issuing long, detailed plans of the day, conducting mast for requests, or minutely inspecting every area of the ship, Carney was a whirlwind of purposeful activity. When I knew more about the navy, I realized that he was performing, with dedication and zeal, the function of an executive officer, which is to be the captain's "designated S.O.B."

But I did not known that then. In my immature and dramatic fancy, Carney was a hot-eyed Calvin of execs, a burn-em-at-the-stake Savonarola, and only strict orthodoxy could save one from a possible inquisition. If all the *California* crew had been as afraid of Carney as I was, his duties would have been very much easier. (My comments about Admiral Carney are based upon youthful memories of him as Commander Carney in the *California*, and should be read in that light. He went on to a truly distinguished career, retiring as a four-star admiral from the navy's highest post, Chief of Naval Operations, in 1955.)

It did not take my radio gang buddies long to discover the dubious pleasures of Bremerton, which were two in number: beer taverns that did not check IDs, and houses of prostitution that catered to sailors.

Within stumbling distance of the navy yard's main gate was the first and most famous bar: the Olympic Tavern. It was known throughout the fleet as Madison Square Garden West. The Olympic advertised itself as the place "where friends and shipmates meet." Anyone who stepped inside those thick, red-brick walls was well advised not to meet his friends and shipmates there but to bring them with him.

The houses of joy were in the usual second floor walk-ups in downtown Bremerton. The navy took a pragmatic interest in these "hotels": the premises were inspected regularly by the shore patrol, and the employees were checked by the ships' doctors.

Most of my liberty hours were spent sightseeing in Seattle. I had the good fortune to make several round trips on what was then the most famous ferry in the world: the M.V. *Kalakala.*

Her name (pronounced *kah-lock-ah-lah*) was Chinook Indian for "flying bird," in tribute to her aerodynamic superstructure, which had been designed by the Boeing Aircraft Company of Seattle. Opinions differed as to her appearance. To some, she looked a little like a floating version of a navy blimp. To others, she resembled, from the stern, a whale with its mouth agape. To me, she looked like a gigantic 1936 Airflo Chrysler, albeit with a blunter nose.

The *Kalakala* was advertised as "the world's first streamlined ferry." She was painted a shiny aluminum, with a green trim between her main (auto) deck and waterline. She had two spacious passenger decks, comfortably equipped with brown, synthetic-leather benches. The fourth, or bridge, deck was fitted with a small wheelhouse that rather resembled a pilot's cockpit with portholes. She made the Bremerton-Seattle run of 15.33 statute miles (13.5 nautical miles) in fifty-five minutes, more or less; the fare was forty-five cents.

Boarding the *Kalakala* from the rickety wooden terminal around the corner from the navy yard gate (still in use and and even more rickety today), I would head for the upper passenger deck, which was recessed to provide a full-service dining room overlooking a broad, elliptical-shaped promenade.

If I had been paid recently, I took a booth and for the price of a twenty-five-cent Olympia beer, with a ten-cent tip for the waitress, enjoyed what is surely one of the most beautiful waterborne routes in the world. If funds were short, I had the same view from a stool at the double-horseshoe lunch counter over a five-cent mug of coffee, with endless free refills.

Powered by a ten-cylinder, 3,000-hp direct-reversing diesel, the *Kalakala* made her way at sixteen knots from Sinclair Inlet into Port Orchard waterway, rounded Point Glover, and headed south-southeast down tortuous Rich Passage. The lush foliage that surrounded us grew right down to the water's edge. Here and there man announced his presence with a small-boat dock fronting a white Cape Cod cottage with its typical gable roof and massive fireplace chimney. Alongside, a solitary gull flapped escort, hoping for a handout. Soon it would disappear, only to be replaced by another.

Overhead were the ever-present clouds of Puget Sound, sometimes thin, high, and fleecy, sometimes dark, low, and threatening. Often they were daubed across the sky in every shade of gray and pink and salmon and coral, as if some color-intoxicated Titan, Vincent Van Gogh a thousand times larger than life, had applied his paintbrushes to the canvas of the heavens.

Twenty-five minutes out, I would catch my first glimpse of a magnificently irregular conical shape standing in splendid isolation to the south, its snow-coverd flanks glowing faintly pink in sunlight. A halo of clouds usually clustered around its 14,410-foot summit. Like Diamond Head and Haleakala, Mount Rainier was a volcano. But it evoked a different feeling: remote grandeur rather than awe tinged with vague anxiety. I could not imagine this rose-tinted mountain harboring a bloodthirsty diety like Pele, the Hawaiian fire goddess.

Soon the *Kalakala* would make a lazy left rudder and the skyline of Seattle would come into partial view. Spread out across its hills, Seattle looked more than a little like San Francisco. But the streets did not climb quite so precipitously, the buildings were brick red rather than glistening white, and the harbor lacked the dramatic sweep of San Francisco Bay.

The full panorama of Seattle's skyline was unfortunately obscured by a low-lying peninsula to starboard, and by the time Duwamish Head had been turned, the impact was lost. But after we passed a red channel buoy and turned toward the brick campanile of venerable Colman Dock amidst a swarm of negligently piloted sailboats, the city began to look more promising.

Seattle might not be another San Francisco, but it had its own charms. For one thing, I had been told, the natives were friendly to sailors. And so they proved. As I explored the city, from the waterfront across its seven hills to Lake Washington and Bellevue on the opposite shore, they were always willing to give directions and advice, and often insisted on taking me where I was going. Seattle, I decided, was a fine city indeed.

I returned early one evening to Bremerton and found my buddies in the Anchor Cafe, across First Street from the ferry terminal. They had been drinking draft beer for some hours and were in boisterous good humor.

One radioman soon made a motion to adjourn to one of the hotels. With some reluctance, I agreed to keep them company. From what I had seen in San Diego and Honolulu, a little commercial sex went a long way.

The reception room was impersonally furnished in tubular chrome and red Leatherette, much like the Honolulu house I would later visit. While my shipmates were taking care of more basic needs, I made an interesting discovery. For thirty-five cents a sailor could buy a mixed drink—something not available elsewhere in Bremerton, as the public taverns were limited to wine and beer. I fed some coins into the jukebox and relaxed with a highball while I listened to "South of the Border" and "Stardust" and other ballads of love unrequited.

An older woman, obviously the madam, came in. Her short but ample figure was stuffed into a low-cut evening gown. Her hair was platinum blonde; her nose as sharp angled as the seaplane crane on the fantail of the *California*. Despite the hair and her heavy theatrical makeup, she projected a certain raffish and worldly maternity.

"You lonesome, sailor?" she asked, sitting next to me.

"No, ma'am," I said. "Not really."

"Polite, too!" she exclaimed. She looked at my rating badge. "Look, Sparks. If you have any doubts, this is the cleanest place in Bremerton. You know how many cases of VD I've had? You know? Just take a guess."

"Very few?" I guessed.

"None!" she said triumphantly. "Never. Not a goddam one. You go out with those little tarts you meet in the beer joints—those lousy amateurs—

Although this photo was probably taken in the late 1930s, it depicts Bremerton's Navy YMCA much as it looked when the author was there in the winter of 1940–41. The bus near the entrance was the type used by the navy to bring girls to ships' dances at Craven Center. At the far left is the entrance to the rickety ferry terminal, operated in those days by the Puget Sound Navigation Company, better known as "the Black Ball Line." (Photo courtesy of the Bremerton Armed Forces YMCA.)

and see what happens to you. Hah! You're restricted to the ship with the clap!"

"Yes, ma'am," I said. "I'm sure you're right."

She leaned closer, peering up at me. "Maybe you're short of money, Sparks. Well, that's no problem for a sailor-boy in my house. Let me show you something."

She led me to a locked door off the reception room. When she opened it, I could see a sizable cloakroom filled with peacoats.

"You see these?" she demanded. "All hocked by sailors who didn't have the money for one of my beautiful girls.

"That's all you need here, honey. Pick out a girl—just leave your peacoat. Redeem it when you get paid. That's what I do for the United States Navy!"

I acknowledged that she was performing a valuable service for the enlisted men.

"I always look out for my sailor-boys," she said proudly. "Now, you ever need a peacoat, you come here. I sell you one much cheaper than storekeepers on ship do." She indicated a long section of peacoats of all sizes, ages, and cuts. "All peacoats left by sailors who got shipped out. Take your pick!"

I told her I had a peacoat but would certainly keep her services in mind. She put a hand on my arm. Her fingers flashed with diamonds and rubies.

"Ah, Sparks," she said. "Muscles. What kind of work you do as a kid?"

I told her I had chopped wood, pitched hay, picked fruit, milked cows.

"Ah," she said again. "That's why you have hard muscle." She squeezed. "And so young and handsome! Sometimes, sailor, I'm sorry I gave up sex."

Her loss was definitely my gain. My raffish hostess was at least forty-five or fifty years old. But I only smiled—bashfully, I am sure—and thanked her for the compliments. It was gratifying to know one could buy a drink here in this navy-sanctioned establishment, and that peacoats were an accepted medium of exchange.

All in one day, I had ridden the *Kalakala* and explored the world of the demimonde with one of them who was a classic in her own way. It had been a fine liberty.

Sailors departing the Puget Sound Navy Yard on liberty were immediately faced with a clear-cut choice. They could turn left, past the Olympic Tavern to the bars and brothels of Bremerton. Or they could turn right, toward the ferry terminal and Seattle. The latter course took them past the entrance to the five-story Navy YMCA.

From this strategic location, the "Y" waged a valiant if often losing battle against the pervasive forces of evil. One of their principal weapons in this uneven struggle was economic. Many of the sailors were only a few weeks or months out of boot camp. Their twenty-one or thirty-six dollars a month would finance few "beer busts" at Craven Center, even at ten cents a bottle, and far fewer in Bremerton or Seattle, where prices were double that or more.

Seamen from the USS *Mississippi* and one lone civilian, probably a shipyard worker, at the entrance to the Bremerton YMCA, one short block from the main gate of Puget Sound Navy Yard. William E. Hanrahan, third from left, has identified this as a 1937 photo. (Photo courtesy of the Bremerton Armed Forces YMCA.)

This 1942 photo shows the lobby of the Bremerton "Y" exactly as it looked when the author was there—except that it is not elbow to elbow with seamen seconds and firemen thirds who preferred to sleep in hard chairs or on the lobby floor rather than return to their poorly heated and ventilated ships. (Photo courtesy of the Bremerton Armed Forces YMCA.)

Most of the entertainment was free at the "Y," which served as a sort of zone of transition between the disciplined military and chaotic civilian worlds. There, with only the occasional presence of the shore patrol a distraction, the seamen seconds and firemen thirds could play pool; work out in the gym; take a swim in the basement pool; write letters home; enter a checkers or chess tournament; or meet a young lady (well chaperoned, to be sure) at a square dance.

Only a few men chose to sing in a chorus or attend Bible class. Religion was not a major factor in navy life. In the 1936 Annual Report of the Bremerton YMCA, it was recorded that 74,900 men used the physical department facilities. Another 29,771 used the pool room. The number attending "religious meetings and Bible classes" was 6,020.

Outside the navy yard gates and across the sound in Seattle, the streets were alive and colorful with shoppers. The external threats from Germany and Japan managed to do what no amount of New Deal "pump priming" had been able to accomplish: restore a measure of prosperity. The United States was, however slowly and reluctantly, activating its massive industrial plant for war, and that meant employment and payrolls for long-idle or underemployed civilians. The navy yard had swollen to more than 10,000 workers; at the Boeing Plant in Seattle, 7,500 men were working day and night, building B-17 Flying Fortress bombers for the U.S. Air Corps and for the hard-pressed British.

The Christmas advertisements in the Seattle *Post-Intelligencer* reflected a buoyant optimism. Newly affluent shipyard and aircraft workers could buy a double-breasted wool Mackinaw windbreaker for $4.98 or a ten-tube Silvertone console, with five broadcast bands, nine push buttons, and an electric tuning eye, for $74.95. A three-piece bedroom set was only $59.88. A Pontiac Streamliner Torpedo Six sedan coupe sold for $923. And for $3,850, one could invest in a five-room brick home in Seattle's Jefferson Park area.

But all this getting and spending, all the jollity of the advertisements, the fake Santa Clauses in the department stores, the Christmas music on the radio, the uniformed Salvation Army enlistees earnestly shaking their tambourines beside their cast-iron contribution pots, all of this had very little to do with the *California*, moored port side to Pier 6, Berth D, in seven fathoms of water. As the biggest holiday of the year approached, no glimmer of the Christmas spirit was evident to me. We existed in a kind of physical and emotional vacuum, isolated from the cares and joys of the civilian world.

I might have believed to this day that all battleships observed Christmas in such a cavalier fashion but for a visit to the *Tennessee*, our sister ship, which was moored to the adjacent Pier 5. Daley knew a radioman in Battleship Forty-three, and I was happy to go along.

Crossing the after brow to the quarterdeck, everything looked precisely the same as in the *California*, including the two-tone mainmast. The customary blue command flag was flying from the main. Yet I was im-

The USS *Tennessee*, the *California*'s sister ship, at Puget Sound Navy Yard in the 1920s. Note the range clock mounted on the foremast and bearing scales painted on number two turret (duplicated on the mainmast and number three turret). These devices, adopted from Royal Navy practice in World War I, enabled ships directly ahead and astern to determine the range and bearing at which the *Tennessee* was firing. (Photo courtesy of the Puget Sound Naval Shipyard.)

mediately aware of a change in ambience. It began with the salute to the JOOW. He actually seemed pleased to have two enlisted visitors from the *California*.

From the quarterdeck, I could, with a little fumbling, have found my way blindfolded to any of my accustomed spaces: the gedunk stand and my mess table on the main deck; up through the wide midships hatch to the casemates and the boat deck; down a starboard ladder to the petty officers' head on the second deck; down again to my berthing compartment, radio two and radio three on the third deck; and yet again, under that sinister smooth hatch cover, to main radio on the first platform.

But the *Tennessee* was not an exact replica of the *California*. She was the only Big Five battleship that still used hammocks rather than bunks. With these primitive sleeping rigs, the simple act of getting in or out of bed became a complicated procedure. At night, the hammock and its enclosed mattress, blankets, and pillow had to be slung from hooks on the deck beams. Then one had to get into the thing: a problem analogous to boarding an Eskimo kayak in rough water. The next morning at up hammocks the stiff canvas had to be unslung, rolled and lashed with seven marlin hitches, and stowed away in the hammock nettings. Certain old-timers—notably chiefs who had advanced to the privilege of innerspring mattresses—swore that nothing in the world was as comfortable as a hammock, but I never saw a CPO request one for sleeping.

(While hammocks had the undeniable virtue of accommodating themselves to the roll of the ship, they posed a real hazard for sleepers who tossed and turned. One night in the *Louisville*, en route from Pearl Harbor to San Francisco in the spring of 1942, I turned the wrong way and was promptly deposited onto the steel deck five feet below. Fortunately, I was merely stunned; no bones were broken. I learned that the only safe sleeping position in a hammock is on one's back.)

The *Tennessee* was different from the *California* in other, more subtle ways. As we moved along the main-deck passageway, past the crew's reception room inboard and the barber shop, post office, and gunnery office outboard, I noticed that the bulkheads and overheads were painted pale green, not the stark white of the *California*'s crew spaces. At about frame 60, athwartships from the ship's service store and fountain, stood a large, lavishly ornamented and gaily lighted Christmas tree. Daley and I paused to stare at it in some astonishment. A real Yule tree on a navy battleship!

Making our way to the radio shack, we met the usual number of crewmen. A few of them actually smiled and said hello. What kind of a battleship was this? We got the same reception in radio one, a duplicate of main radio in the *California* except for the green paint and a few alterations in the arrangement of the receivers and the communication office (the

After her launching ceremony at the Mare Island Navy Yard on 20 November 1919, the new battleship *California* enters the water—and keeps right on going toward the Vallejo side of the narrow channel. She stopped only after taking out twenty-five or thirty feet of the ferry dock. The legend grew until yard workers who rode on this shortest of her cruises swore that she went halfway up Georgia Street. Battleship Forty-four was christened by Mrs. Barbara Zane, daughter of California Governor William D. Stephens. (Photo courtesy of the Naval Institute Collection.)

result, no doubt, of many navy yard refits and overhauls.) Lacking was the stomach-knotting tension that was always felt in the *California*.

As we exchanged scuttlebutt with the watch standers and idlers, I realized that I was getting much better training under the direction of Chief Reeves than I would in our sister ship, but the latter would be a much more pleasant place in which to serve. For "the *California* was cold and austere, resonant with the measured tread of three- and four-star admirals long gone from her decks, taut with the overwhelming presence of the current Commander Battle Force and his ambitious and driven staff. The *Tennessee*, rating only a division flag (and that nominal, being in Battle[ship] Division Two with the *California*), had no such burden of command: she was relaxed, for a battlewagon, and happy-go-lucky."*

For close to nineteen years, the *California* and *Tennessee* had been nearly inseparable companions, despite the fact that they began life a continent apart. The *California* was laid down at Mare Island Navy Yard on

*Theodore C. Mason, U.S. Naval Institute *Proceedings*, vol. 106, no. 4, April 1980, p. 85.

25 October 1916, launched 20 November 1919, and commissioned 10 August 1921. (Her commissioning officer was Captain Edward L. Beach, USN, commandant of the yard and father of the distinguished Captain Edward L. Beach of submarine and novel fame.) The *Tennessee's* keel was laid at New York Navy Yard on 14 May 1917. She was launched 30 April 1919 and commissioned 3 January 1920. They were the first American battleships built with a post-Jutland hull design featuring extensive underwater protection, secondary batteries located on the upper deck level, and fire-control systems for both main and secondary batteries.

After her shakedown cruise and official trials, the *Tennessee* transited the Panama Canal and joined the other U.S. Fleet battleships at their home port of San Pedro on 17 June 1921.

Early in 1922, the *California* arrived to become flagship of the U.S. Battle Fleet (renamed the Battle Force in 1931). From then until a certain December day in 1941, the sisters operated together, separated only for periodic navy yard overhauls.

Together, they made the cruise of the Battle Fleet to Australia and New Zealand via Samoa in 1925. Together, they passed through the Panama Canal in February 1930 to participate in Fleet Problems X and XI in the

This rare photo, dated August 1920, shows the *California* under construction at Mare Island. The only dreadnought ever built on the West Coast was placed in commission on 10 August 1921 by Captain Edward L. Beach, commandant of the navy yard. After elaborate commissioning ceremonies, including full-dress ship, the commandant turned her over to Captain Henry J. Ziegmeier, her first CO. (Photo courtesy of the Naval Institute Collection.)

Caribbean and in the presidential review of the fleet off Chesapeake Capes, Virginia, on 20 May 1930.

Back in San Pedro, they lay at anchor a cable length apart off the breakwater all through the 1930s, steaming out to participate in joint army-navy and tactical exercises, fleet problems, and good-will cruises up and down the Pacific Coast from Alaska to Panama.

I thought about those "piping days" as I prepared to leave the *Tennessee*'s quarterdeck and wished I had been old enough to be part of them. Unknown to me, the sisters would be together again at Pearl Harbor on a "day of infamy" one year later. And they would still be together on 25 October 1944, when they and the battleships *West Virginia, Maryland, Mississippi,* and *Pennsylvania* capped the Japanese "T" at Surigao Strait in the Philippines and sank the battleship *Yamashiro* in a few minutes of concentrated gunfire. It was the last battle-line engagement in navy history, and it seems altogether fitting that these two valiant and sturdy sisters were active participants.

We must not leave the *Tennessee* without correcting a notion that was nearly universal in the fleet in 1940. As a crewman, I was told that the *California* began life as the *Tennessee* at Mare Island and the *Tennessee* as the *California* in New York. During construction the Department of the Navy obligingly switched the names so that the state of California could have the honor of building its own battleship.

I had no reason to question this tale and, until very recently, accepted it as gospel. Seamen who had scraped the bottom of the *California* told me that her anchor, armor belt, and keel were emblazoned with the inscription *U.S.S. Tennessee BB 43.*

As if to confirm the story beyond doubt, during the attack on Pearl Harbor a bomb penetrated the five-inch crown armor of the *Tennessee*'s number three turret and exploded with a low-order detonation. A fragment from the turret hit the marine barracks on Ford Island and passed through the bunk of one of the marines. The fragment had the name *California* painted on it. Since the *California* suffered no turret damage, the piece of steel must have come from her sister ship.

Unfortunately the legend, despite this impressive evidence, is only half true. The *California* did indeed change names—but not with the *Tennessee*. The ship she changed names with was the *New Mexico*. The change apparently took place as a result of an intensive lobbying campaign in Washington by California politicians, Vallejo officials, and civic leaders.*

---

*Mr. Richard T. Speer (Ships' Histories Branch, Naval Historical Center, Washington Navy Yard) gave me the facts in a letter dated 4 June 1981.

The Navy appropriation act of 30 June 1914 authorized construction of Battleships 40–42. On 30 July of that year the Secretary of the Navy assigned the names CALIFORNIA, MISSISSIPPI, and IDAHO, respectively, to them. CALIFORNIA (BB 40)

A reasonable explanation for the fiction of the *California-Tennessee* name inversion is that it was common navy practice to interchange parts and fittings among sister ships. In the absence of any documentation, it is no wonder that the *California* plank owners believed the name change had taken place within the class.

This scuttlebutt was passed down from one generation to another of crewmen. Oral history, like firearms or firewater, must be handled with caution.

The day before Christmas, Daley came looking for me. He was even more sober faced than usual.

"I just got my orders," he explained. "Receiving Station Puget Sound, FFT East Coast for new construction. Leaving day after tomorrow."

"Damn, I'm sorry to hear that," I said. We were good friends, and I would miss him. "What ship?"

"The *Fuller*. Old passenger tub the navy is converting to a transport." He smiled faintly. "Nothing's too good for our soldiers, you know."

"An AP might be pretty good duty," I said consolingly. "Less spit and polish. Should be plenty of places to hide out and write. It's just too damn bad you'll have to leave Mary Jane."

"I'll probably lie around the receiving station for a month or so. I'd ask you to keep an eye on her, but I hear you guys are shoving off any day now."

In view of the circumstances, that was all right with me. If Daley were to leave Bremerton before I did, my knotty problem concerning the ethics of friendship would have to be wrestled with all over again.

"There's going to be a midnight high mass on the *Enterprise* tonight," Daley continued. "How about going with me?"

"I'd be glad to," I said. "But hell, I'm not a Catholic."

"That doesn't matter. We like to get Christian Scientists and other unbaptized infidels. Makes it easier to convert them." Laughing, I promptly agreed to attend my first Catholic service.

Daley and I stopped at dockside that evening to examine the *Enterprise* through the critical eyes of battleship sailors. She looked too big and clumsy to be a ship and too small to be a mobile airfield. Below her flight

---

was ordered from New York Navy Yard, while MISSISSIPPI (BB 41) was awarded to Newport News Shipbuilding and IDAHO (BB 42) was given to New York Shipbuilding of Camden, New Jersey. The next appropriation act, signed on 3 March 1915, provided for two battleships (43 and 44). BB 43 was assigned to the New York Navy Yard and named TENNESSEE. BB 44 was ordered from Mare Island Navy Yard on 28 December 1915. On 22 March 1916 BB 40 was renamed NEW MEXICO; the armored cruiser CALIFORNIA was renamed SAN DIEGO, and the new BB 44 was given the name CALIFORNIA.

(The keel of the *New Mexico* [ex-*California*] was laid down 14 October 1915; she was commissioned 20 May 1918.)

deck, the tip of a bow projected out timidly. That much of her looked like a ship. Otherwise, she resembled a naval architect's mistake that had been compensated for by slicing off everything above the main deck and slapping an ill-fitting slab of runway atop the remains. Besides, she was lopsided: her funnels and attenuated superstructure had been positioned, as if by afterthought, along her starboard deck edge.

No matter that she could handle up to one hundred planes and that her 120,000-hp geared turbines gave her a speed of thirty-four knots. She had very little armor protection; her only armament was eight of the new 5-inch 38-caliber dual-purpose guns and a number of the new 1.1-inch pom-poms. I remembered how she had looked at sea, pitching and rolling arthritically, with a seemingly perpetual starboard list. If she were an enemy and the *California* ever got within 30,000 yards, we could blow her out of the water with a couple of salvoes. How could we miss such a huge, slow-turning target?

"How would you like to serve in one of these things?" I asked Daley.

He gave me a look of mock horror. "I'll take the *Fuller*," he said. "Hell, I'd even take the *Nitro!*"

Since ammunition ships were considered the worst and most dangerous duty in the fleet, that was a strong statement, but I agreed with him.

"Let's look at the bright side," he added. "This seagoing bathtub is safe at the dock. That should make her ideal for a midnight mass."

The 800-foot-long hangar deck of the *Enterprise* had been strung with Christmas lights and converted for church services with many rows of mess benches. At the forward end stood a large, portable altar with its six candles, tabernacle, white altar cloths, and white antependium decorated with silver crosses. Over the altar, a large crucifix had been suspended from the overhead. I estimated there were some 3,000 officers and men present, an impressive sight in their dress-blue uniforms. A sprinkling of officers' wives and other guests added a touch of femininity and color.

The area nearest the altar had been roped off. The rows of wardroom chairs were occupied by officers and their ladies.

I nudged Daley. "Officers' country up forward. Do they think they're going to the head of the line at Saint Peter's gate?"

He considered the question for a moment. "That won't be any problem," he quipped. "Most of the officers I've known are going in the opposite direction."

The mass was celebrated by a priest and two assistants adorned in long, flowing white chasubles, richly ornamented with silver and gold embroidery. The altar boys and choir of enlisted men were attired in plainer white vestments.

Just before the service began, I suddenly had an alarming thought. "M.G.! How will I know what to do?"

"It's just like dress ship," Daley whispered. "Follow my movements."

The elaborate four-part mass was chanted and sung in Latin. As I mimicked Daley in making the sign of the cross, genuflecting, and kneeling, I strained to pick out words and phrases remembered from my high-school Latin courses. But Caesar's *veni, vidi, vici* and his *Commentaries on the Gallic War* were of little help.

I remembered something my great-uncle had told me: "There are many paths up the mountain toward the ultimate truth." Since this one had stood the test of nearly twenty centuries, it obviously had much to recommend it.

I was glad I had exposed myself to a new and moving experience on this Christmas morning. It was a felicitous way to bid farewell to my friend, and farewell to Bremerton.

# CHAPTER 7

# Bad Luck at the Breakwater: A Couple of Jonahs?

At 0900 on the last day of 1940, residents of Long Beach, Wilmington, and San Pedro were treated to an impressive sight: five mighty, gray battleships anchored in column inside the breakwater, and the entire crew of the leading dreadnought mustered on the quarterdeck in dress blues. Nearby were two huge aircraft carriers, the *Lexington* and *Saratoga* (identical twins except for the vertical funnel strip on the *Sara*), three cruisers, four destroyers, and a number of fleet auxiliaries.

I hoped the civilians were enjoying the view. Being one of the crew of that leading dreadnought, I wasn't. My shipmates and I were drawn up in division parade for inspection and a change-of-command ceremony.

Change of command is a formal rite with quasi-religious overtones, little changed from the days of John Paul Jones. The officers, chief petty officers, and crew were formed in ranks around three sides of a lectern set up abaft number four turret. A row of wardroom chairs curved around behind the lectern to accommodate guests and the braid-encrusted official party. After an invocation by our distinguished-looking Catholic chaplain, Commander William A. Maguire, Admiral Snyder made a few terse remarks. I had seen him before at captain's inspection: a short stocky man with a small hard mouth, a long bent nose, and graying hair worn with no sideburns at all, in the 1920s style popularized by Jack Dempsey. Commander Carney, that stern archetype of an executive officer, read his orders and made a few even terser remarks. Then it was our new commanding officer's turn.

Although ceremonies of this kind usually attained at best a kind of dull solemnity, the crew did not find this one boring. Far from it. This was their chance to evaluate their new skipper, a matter of the most vital concern. Very few men in civilian life were ever exposed to such a close and searching scrutiny as the crew of a navy combat ship gave its new CO. For it was he who determined whether it would be a happy ship or an unhappy one; a taut ship or a slack one; a ship with esprit de corps or one morose and dispirited. He alone determined whether the food would be palatable, the quarters livable, the discipline harsh or tempered with mercy. Finally, the captain might well determine, by his competence or incompetence, whether any or all of us standing on the quarterdeck that Tuesday morning lived or died. So all eyes were fixed on Captain Joel W. Bunkley as he stood at the lectern to read his orders, and men in the rear ranks craned their necks for a better view.

He was a man of above-medium height, already in his early or middle fifties, I guessed. He wore his cap low across his eyebrows, partially shielding his face. Later, I tried to remember whether I had ever seen him without it and decided I had not, since I couldn't remember what color his hair was and how much of it remained. The cap seemed as much a part of him as his ears. In fact, I couldn't imagine him without his uniform on. I had the fanciful notion that without those gold-trimmed props, he would change magically before my eyes and become a shriveled, grotesque old man.

He had a long, well-structured face, but his eyes were as bleak as two pieces of polished basalt. Although he was darkly tanned, his skin was weathered and blotchy. He did not look healthy to me, and my intuitive diagnosis was correct: during the next year he would be periodically confined to his cabin with vague indispositions.

He read his orders in a cold, formal voice that had an impatient, raspish edge. The nature of their job seemed to make battleship commanders severe and dispassionate figures—this one appeared especially so.

Fisher was standing next to me. We shot each other quick glances. "God damn," we chimed, under our breaths.

At the tall end of our rank, M. G. Johnson, who was sweating out the verdict of another summary court-martial for AOL, looked dismayed. "Jesus Christ!" he said in a stage whisper. "They may throw the frigging key away!"

Bunkley's remarks did nothing to change our first impressions. He seemed to be a by-the-book officer giving a few by-the-book comments that promised drills and more drills; exercises and more exercises; and close observance of navy regulations.

We had no sooner reached our living compartment for the shift into undress blues than the general alarm sounded, followed by the bugle call for assembly.

"Away the fire and rescue party!" bawled the voice from the ship's PA system. Since the radio gang was exempt from this drill, we paid no attention.

A bull session about the new skipper was just getting under way when it was disturbed by a long blast of the siren and the atonal ululation of the warning howlers.

"Collision in the location of the starboard bow!" shouted the boatswain's mate of the watch. We ignored this message, too.

We couldn't ignore the next one. The general alarm bonged. The bugler sounded two calls in rapid succession, the second of which I recognized as double time. The bosun's pipe trilled.

"All hands abandon ship!"

This command was greeted with dismay.

"Where the hell is our abandon-ship station?" we asked each other. No one seemed to know. There was a sprint to main radio to consult the watch, quarter, and station bill. Eventually, we reached the boat deck, where we milled about in hopeless confusion.

Schrader, our leading petty officer, showed up to restore order and direct us to the proper boats.

"The captain sure as hell won't be the last man off this ship," he sniffed scornfully. "The goddam radio gang will!" His remark was prophetic for one radioman.

As we secured from abandon ship, the general alarm sounded again. "General quarters! General quarters!" We raced to our battle stations. At least we knew where they were. After an hour of drills, the crew was paid.

"Who is this horseshit bastard?"

That, in its various permutations, was the question on everyone's lips. We felt insulted by Bunkley's "whip-and-pogey-bait" approach.

"This 'horseshit bastard' is your new CO," one of the senior firsts grinned. Having seen many changes of command, he assured us that the dust created by the new broom would soon settle and things would fall back into their customary routine.

"Meanwhile," he added happily, "you people had better shape up or you'll be shipping out. You might start by finding out where you're supposed to be at general drills."

But we were more concerned about who Bunkley was and where he had come from. No one had ever heard of him, so there was nothing to do but dispatch our operatives to the captain's, exec's, and chaplain's offices, as well as the chief's quarters. The "yeomanettes" in those offices usually knew what was going on, if they could be persuaded, cajoled, or bribed into talking; and, of course, Chief Reeves would have the word.

Rivulets of information soon began trickling down to the third deck. For the past year or so, Bunkley had commanded a naval ammunition depot.

"Ammunition depot? That's the last stop before retirement, isn't it?"

Before that, he had run a destroyer division for a year.

"He probably spent most of his time swinging around a goddam buoy at San Diego."

From 1934 to 1937, he had been on a naval mission to Argentina.

"Three years on his ass in South America! The Argentine Navy? What navy?"

Reaching back to 1933, we learned he had actually been exec of a light cruiser, the *Marblehead* (CL 12).

"The *Marblehead*? It's nothin' but an overgrown four-piper, flush-deck destroyer, makin' the milk run from Cavite to Shanghai."

The only ships Bunkley had ever commanded were destroyers, back in the late 1920s and early 30s.

"Tin cans are okay–but what has that got to do with the battleship navy?"

The only battleships he had ever served in were the *Connecticut* from 1909 to 1912, right after he graduated from the Naval Academy, and the *Wyoming*, where he was assistant gunnery officer in 1915.

"The *Connecticut* is long gone and the *Wyoming* was demilitarized in 1931. Bunkley hasn't served on a battlewagon since before we entered the First World War, for crissakes!"

Now how the hell, we asked ourselves, does a guy this old get from a navy ammo dump to command of a battleship? Not just any battleship but the flagship of the Battle Force! Had the brass hats at the Bureau of Navigation gone Asiatic?*

One old-timer listened to us disparage Bunkley's record with growing exasperation. "What you guys know about the navy could be written on a piece of shit paper—and then you could wipe your ass with it," he growled.

"First of all, command of an ammunition depot is an important job. That's where the research on gunnery is done. Bunkley is a gunnery expert. In other words, he's a member of the gun club."

"Gun club? What the hell's that?"

"Oh, nothin'," he replied with heavy sarcasm. "Just the admirals and four-stripers who run the navy. They're all big-gun men. They don't give a shit about them covered wagons you see out there. They're battleship men. Bunkley's one of 'em. When they got a chance to dump Bemis, they gave him the slot."

"Political connections, huh?"

"If you want to call it that. Just remember, he knows fourteen-inch guns. That's what a battleship is, God damn it, guns. It's just one big floating gun platform."

"He hasn't been near a battleship for twenty-five years!"

---

*Service in the Asiatic Fleet at Chefoo (South China Patrol) or Shanghai (Yangtze River Gunboat Flotilla) was highly prized by some career sailors. A petty officer's pay commanded a life of comparative luxury, complete with houseboy, mistress, and all the forbidden fruits of the Orient. When these sailors reluctantly rejoined the U.S. Fleet (many sporting a wondrous assortment of tattoos and a gold earring in the left ear), their behavior was often considered bizarre or even mentally unbalanced. They had, in the popular navy phrase, "gone Asiatic."

"How much do you think they've changed? It's not like Detroit, where they retool and turn out a new model every year. The twelve-inchers on the *Connecticut* and the *Wyoming* operated exactly the same way as the main battery on this ship. All they've done is increase the elevation so they can fire over the horizon and add planes to spot the fall of shot. And they've improved the director optics and fire-control system a little. The principle's exactly the same."

We didn't know enough about ordnance to argue with him, so we had to fall back on calumny. "How come the old bastard's so horseshit, then?"

"He's Old Navy," was the answer. "He's been away a long time—and he wants you snot-nosed kids to know who's boss. Once that's established, he'll be okay."

Chief Radioman Reinhardt shambled in to supply the final word. A tall bearlike man with an ample beer belly, he loved to lecture young sailors. In truth, he had little else to do. Although he was a flag chief, he was junior to Reeves, who ran everything. Reinhardt spent most of his time drinking coffee in flag radio or the chiefs' quarters, making periodic forays into the two radiomen's compartments in search of an audience.

"I've been telling you goddam gedunk sailors about the real navy," he bellowed at his customary volume. "Now you're gonna find out for yourselves what it's like to serve on a battlewagon, not a goddam yacht!"

We groaned. This was a hell of a way to start a new year!

About this time, I met the first ship's officer who seemed to show any real concern for the welfare of the enlisted men. (Of course, I was judging from the narrow perspective of the ninety-day wonders in C-D Division. In truth, there were a number of fine Annapolis-trained officers in other divisions. One of them was Ensign T. P. McGrath, C-L Division officer, who was highly regarded by his signalmen.)*

The Bunkley broom was making a clean sweep-down, fore and aft, which included assembly of the permanent landing force. In those days, radiomen were expected to "double in brass" as rifle squads, in addition to operating the portable transmitters they would take ashore in any landing on a hostile beach; so I was not surprised that most of an entire watch section of the radio gang was listed.

Fisher, Stammerjohn, and I were on the landing force bill, along with a number of other junior ratings and strikers.

---

*There was good reason for this assessment. Six feet three inches tall and ruggedly handsome, the young signal officer was a natural leader of impressive presence. Thomas Patrick McGrath, USNA '40, was regimental commander and captain of the football team at Annapolis. In early 1942 he went to the submarine service. He was executive officer of the *Pompano* when she was lost with all hands off northeast Honshu in August 1943.

Handsome Joe B. Ross, one of the "shoreside operators" of the radio gang, cuts up in hula costume at Waikiki in late 1940. In his right hand are appropriate liquid refreshments. (Photo courtesy of Vernon L. Luettinger.)

"All the expendables," joked Joe B. Ross, who was not, of course, on the bill. "And what the hell is more expendable than a V-3?"

Joe had recently made his second-class rating and therefore looked down on all radiomen thirds. He was especially condescending to radiomen thirds of the communication reserve.

He was one of the big shoreside operators of the radio gang—or thought he was. Exceedingly handsome and well built, he looked more than a little like movie idol Robert Taylor. The bane of his existence was his receding hairline, which he continually massaged with new lotions and potions brought back from the beach. Brought back, too, were reports of his latest seductions, delivered in loud and boastful detail in his twangy Texas accent.

Ignoring Ross's gibes, we fell in near the crew's reception room in undress blues with leggings, cartridge belts, gas masks, and helmets (still the World War I doughboy type), and struggled to assemble our light-marching-order packs. Springfield service rifles were handed out, the occasion for horseplay, pseudo-threatening gestures with bayonets and rifle butts, and light-hearted speculation about our destination.

"We're going to deploy along Beacon Street and clean out all the drunken merchant marine sailors," one radioman guessed.

"Naw," another replied. "We're going to advance on the Pike and engage the whores in close-order combat."*

"Then sick bay had better stand by to receive casualties!"

Most of my fellow thirds, although my superiors with headsets and telegraph keys, looked comically ineffectual as marines, something like boy scouts playing at war. I decided that their youthful faces, peering out from under the brims of their tin hats, would scarcely frighten an old maid, let alone our potential enemies. The gingerly, amateurish way they handled their pieces made them a greater menace to themselves and their shipmates than they could possibly be to the Germans or Japanese.

Lieutenant H.E. Bernstein, our communication division officer, soon appeared to make a preliminary inspection. He did not look any more Jewish than Commander Carney looked Irish or Admiral Nimitz looked German. All had the faces of men who have transcended their antecedents, a phenomenon that I have often noted to be a by-product of the profession of arms. He was of medium height and rather stout build. He wore his slightly rumpled blue uniform as if it were a business suit purchased off the rack. Always, I visualized him behind a desk, operating in a kind of benign chaos, ashes from a forgotten cigarette spilling across his lapels.

Although not noted for his spit and polish, Lieutenant Bernstein was an Annapolis graduate, class of 1926, and one of the very few Jewish Naval Academy men of that time. A specialist in electronics, he was also one of a very select number of officers who knew anything about our revolutionary CXAM radar. In addition, he was the ship's security officer. As he passed down the line, stopping here and there to ask a question or help with an equipment adjustment, he appeared to be enjoying himself. Usually, he stayed behind the scenes and seldom had any contact with enlisted men. (That role was played by the always-ambitious flag communication officer, who outranked him by a half or full stripe.)

When he got to me, he paused.

"Let's see. You're Mason, aren't you?"

I was surprised he knew my name. "Yes, sir."

"One of our reserves. I've had good reports about the three of you. How are you getting along?"

I already knew that when most officers asked that question, they didn't want a truthful answer. "Fine, sir," was expected, and "Fine, sir," is what

---

*Actually, we landed and set up our radio gear on Terminal Island. Less than a year later, we would be digging in for real on an island named Ford.

they got. But there was something different about Bernstein. He looked
kind and genuinely solicitous, the sort of man one could trust.

So I surprised myself by answering, "It's a little rough, sir. I have a lot to
learn."

Bernstein smiled. "We all do. Any special problems?"

Now I had got myself into an embarrassing position, but I had to
continue. "No, sir. It's just that I wasn't a ham operator before I volunteered, so I'm having to catch up now."

Again the smile. He reached over and made a minor adjustment to one
suspender of my pack, a comradely gesture. "I'm sure you'll do all right,
Mason. You seem like an intelligent young man." He started to move on,
then stopped. "It may be a little rough now," he said. "But it's for a good
cause. Just remember that."

Our brief conversation left me with a warm glow. I was not just a
number, not just a third-class radioman who had yet to master his specialty:
I was a person of value. Bernstein and I knew that duty in the *California*
was tough; Bernstein and I knew that it was for a good cause: temporarily
sacrificing some freedom to preserve a larger freedom. It hardly occurred
to me then that Lieutenant Henry E. Bernstein, far more than I, knew just
how important a cause it was.

Our new CO continued to serve notice on us "snot-nosed kids" that a
battleship skipper's authority was as near absolute as a government of free
men could permit. Despite the fact that we had had a personnel inspection
less than a week before at the change-of-command ceremony, he ordered
another one for Saturday, 4 January. At captain's mast on Friday, a luckless
seaman first was given a general court-martial on charges that did not seem
to warrant such grave action (a general court-martial could result in many
years at hard labor). The log of 3 January 1941 accused him of "bringing
discredit upon the Naval uniform, using profane and obscene language to
Shore Patrol," being "drunk and disorderly," and "resisting arrest."

On 8 January, another seaman first who was AOL for three hours and
forty-five minutes was deprived of sixteen liberties (facing, thereby, a
confinement to the ship of at least two months). A fireman third who had
the wit to return only two hours and forty minutes AOL lost twelve
liberties.

A week later, for various offenses, Bunkley passed out a summary
court-martial, a deck court, five days' solitary confinement on bread and
water, and deprivation of ten liberties. Two third-class mess attendants
who were reported for fighting were treated more benevolently: three
days' solitary on B & W.

As we were returning to our anchorage from gunnery exercises on the
late evening of 9 January, there occurred one of those inexplicable and
unforeseeable personnel failures that cause captains to gulp antacids and

turn gray prematurely. Just as we were rounding the breakwater, the ship suddenly lost all power. It was restored in time to avoid drifting onto the rocks, but we had had, all hands agreed, a very close call. The responsibility was soon fixed on the black gang.

Naturally, the snipes were reluctant to discuss a breakdown they could not blame on the yardbirds. They spoke vaguely of a shot of water that had stripped the blades on one of the giant steam turbines. Apparently, a watertender had failed to keep the feedwater at the proper level in his glass.

In any event, the foul up caused an unscheduled return to Bremerton that put the ship *hors de combat* for nearly three months. The Battle Force flag had to be transferred, with repercussions that echoed all the way to Puget Sound and Pearl Harbor. Not the least of these were a great increase in radio traffic and a vast consumption of correspondence forms and typewriter ribbons.

Gloomily, the two dozen men of the flag radio complement lashed their bags and hammocks for transfer to the *Boise* (CL 47), which would deliver them to the relief flagship *New Mexico*, already at Pearl. We bade them a gleeful farewell, promised to toast them in beer at the Music Hall or the Garden of Allah, or perhaps in stronger stuff at one of the Seattle bottle clubs—and promptly took over the choice lower bunks and upper lockers they had vacated.

Early on the afternoon of 20 January, the marine guard, the ship's band, and eight side boys were paraded. Admiral Snyder and his staff officers were attended at the starboard quarterdeck by Captain Bunkley and the ship's senior officers. The guard presented arms, the band sounded off with "The Admiral's March," and Snyder was piped over the side and into his waiting barge to the shrill treble of the boatswain's pipe. Within a few minutes, his four-star flag fluttered from the *Maryland*'s main. Smartly, the *California*'s signalmen ran up the thin ribbon of the seven-starred commission pennant and hauled down the ComBatFor flag to a general feeling of relief throughout the ship. For a little while we were free of the twenty officers and eighty or so men who so greatly reduced the habitability of the ship and so sharply increased the level of tension and anxiety.

An hour later I watched the *Maryland*, *Tennessee*, *West Virginia*, and *Colorado* up anchor and pass the breakwater in single file, silent and gray and wraithlike against the smoky southern horizon. Four of the Big Five were returning to duty in Hawaii, where our commander in chief would very soon lose his job over the issue of basing them there. (Admiral Richardson was ordered relieved on 31 January 1941. His successor was Admiral Husband E. Kimmel.) But the fifth battleship got under way for the far more congenial waters of Puget Sound at 1552; and this time our captain officially took the conn.

We steamed up the coast at fourteen knots, working up to fifteen. Along the magnificent northern California–Oregon littoral, rain squalls, haze, and light fog partially obscured the steep escarpments, isolated sandy coves, and rocky peninsulas knifing into the Pacific. Marooned offshore were many jagged "sea stacks": the beleaguered rear guard of a coastline that was retreating inch by grudging inch against an amphibious assault that had begun before there were eyes to observe it. Born in the seas off Japan and gathering momentum over 6,000 miles of unbroken water, the giant swells had been furiously and triumphantly hurling themselves against this granite citadel for millennia.

The atmosphere in the radio shack was relaxed and expectant. I had been promoted to the weather circuit, and was no longer a fumbling scrub but a first-stringer of sorts. To be sure, it was deadly dull copying five-digit code groups, but the broadcasts didn't run continuously, and I could hang the earphones around my neck and join in the general merriment.

While my watchmates were debating the merits of various Seattle bars, one seagoing thinker (with which radio gangs were always plentifully supplied) interjected a sobering thought.

"This new skipper isn't aboard more than a week before we wreck a turbine, or something, and have to go limping back to Bremerton. You guys ever think he's bringing us bad luck?"

Any mention of the ship's captain—especially one as controversial as Bunkley was proving to be—was certain to attract an audience.

"If it wasn't for that foul up," one liberty hound scoffed, "we'd be taking a taxi to a Honolulu cat-house instead of riding the *Kalakala* to Seattle. Now what kind of bad luck is that?"

Ignoring this obvious truth, the philosopher second class—a swarthy fellow with keen, thoughtful brown eyes—responded with his own question.

"Any of you know who our engineering officer is?"

No one did. When it came to other divisions, we didn't know one officer from another.

"I'll tell you. His name is Naquin. Lieutenant Commander Oliver F. Naquin. That ring any bells?"

The name was tantalizingly familiar but just out of reach. The philosopher looked from face to blank face, enjoying our ignorance.

"Well, he's the guy who lost the *Squalus*." With heavy emphasis he added, "On her shakedown cruise."

That intelligence set off a buzz of conversation. Everyone had heard of the submarine *Squalus* (SS 192). Its loss in May 1939 had been front-page news for days. The new sub, commanded by Naquin, sank in 243 feet of water off Portsmouth, New Hampshire, when a main air induction line failed to close during a test dive. Twenty-six men drowned; the other thirty-three were saved by means of a McCann rescue chamber. The press was there to cover every heart-stopping moment of the rescue in vivid if wildly inaccurate prose.

"What's wrong with having a hero aboard?" one radioman demanded.

"The newspapers made him a hero," our guru replied. "The navy doesn't agree. He lost a brand-new sub on its shakedown. He won't get another seagoing command until hell freezes over. And he'll never get near a pigboat again. That's why they shipped him here."

I got into the conversation for the first time. "So he lost the *Squalus*. That doesn't mean it was his fault."

"It doesn't matter whose fault it was, Mason," the second class explained. "The captain takes the rap. That's navy tradition. What bothers me is that he's an unlucky officer. First, the *Squalus*. Then, his black gang screws up and we damn near pile into the breakwater."

Now that he had our undivided attention, he continued eagerly. "You guys figure it out for yourselves. We got an engineering officer who's already lost one boat and damn near loses this one. We got an overage tin-can skipper with political connections in Washington who barely makes it back to port his second time out on a real ship. We got us two Jonahs on the Prune Barge!"

There were nods and murmurs of agreement, but the liberty hound didn't join in. "I don't believe in that bullshit," he jeered. "Soon as I hit the ferry, I'm gonna pour an Oley to those two Jonahs, as you call 'em, for gettin' us back to Bremerton."

Our philosopher was unmoved. "Just wait and see," was his confident prediction.

I didn't believe in such superstitions, either. I didn't believe in Davy Jones or the *Flying Dutchman* or mermaids with long black hair and porpoise's tails. I didn't believe in St. Elmo's fire or sea monsters or fortune tellers and soothsayers. I didn't believe in precognition, I didn't believe that Friday was an unlucky day—and I didn't believe in Jonahs.

Not then I didn't.

We came into Puget Sound on a chill morning of lowering clouds and light rain and anchored north of Blake Island. A couple of hours later a pilot came aboard to conn us through Rich Passage and the Port Orchard waterway. As we approached the navy yard, the crew of the saluting battery was called to quarters and we fired a thirteen-gun salute to the commandant of the Thirteenth Naval District, a rear admiral. Promptly, the return salute of seven guns came from the roof of the administration building. Two oceangoing tugs secured to our starboard quarter and bow and eased us into Berth B, Pier 5. And there we were to stay for the next two-and-a-half months.

With our main engines and boilers secured and the ship receiving electricity, fresh water, compressed air, and other essential services from the dockside umbilical cords, we fell into a typical navy yard routine. Lucky Michael reported in from Long Beach with the ship's service truck. The yardbirds straggled aboard and descended into the deep engineering spaces. (Oddly, no one in the radio gang tried to cultivate one of them so we could find out what really happened on the evening of 9 January. On my part, at least, it was an act of service arrogance for which I later awarded myself a dozen lashes.) Men were detached from the ship for temporary duty with the shore patrol and the military guard patrol, both based at the Bremerton police station. On weekends, the SP force was beefed up with additional line petty officers from the ship.

The movie and dance patrols at Craven Center resumed. Other radiomen were selected for this easy duty, including Fisher and Vernon I. Luettinger, a quiet, abstemious third class who was the bantamweight on the boxing team, but I was ignored. I wondered if Ensign Striker had had anything to do with that.

"Your trouble, Mason, is that you're too goddam GI," Fisher chortled. "Don't you know better than to stop an admiral over a stupid thing like a pass?"

"I was just following orders," I said, with the stock rationalization of bureaucrats (and war criminals). Uncomfortable at having to fall back to this position, I awaited my chance for suitable retribution.

I had become friendly with a third-class radioman named Jerry Litwak, who had reported aboard on our previous trip to Bremerton. In his early twenties and already balding, he was keenly intelligent, had a fondness for Artie Shaw records, and a cutting, satirical sense of humor that made him rather a loner in the radio gang. Idling in main radio one evening, we composed a long, mildly bawdy poetic account of Fisher's shoreside activities. I would write the first line of a couplet, and Jerry would supply the punch line that completed it. This was accompanied by howls of laughter at our own brilliance. A sample follows:

> Fisher went into Bremerton hunting a squack;*
> Two weeks went by and he still wasn't back;
> For innocent Fisher was horribly cursed,
> And for cherry and lime Cokes acquired a thirst.
> 'Twas an ill-fated day when he entered those rooms,
> And his mind was bewildered by stale beer fumes. . . .

So pleased were we that we promptly made a number of carbon copies of our epic, which we distributed among the radio gang.

Fisher was not amused.

Shorn of our flag personnel, and with most radio circuits secured, the

*"Squack" was navy slang for a native girl of the Pacific, and by extension, any young lady considered to have loose morals.

radio gang adopted its own carefree procedure of few watches and many liberties. A land line was set up in radio one, at the watch supervisor's desk, for fast communication between ship and yard. The telegraph sounder clicked and clacked away in a mysterious gibberish that I totally failed to decipher as Morse code. Only the old-timers could copy it—and not even all of them. I had no intention of learning to use the land line. It did not seem to me to have any future, and my prediction of its demise proved accurate.

A first class and two thirds—including Stammerjohn—were detached for temporary duty at radio station NPC Puget Sound. Stammerjohn was elated, for at last he would have a chance to control a live circuit with a heavy traffic load and prove what he could do.

I, too, had my brief hours at NPC. A few days after we moored to Berth 5-B, Schrader, the leading petty officer, came looking for me. "I want you to report to NPC at 2330," he said without preamble. "They're shorthanded and need a third class."

"Tonight?" I asked blankly. "Me?"

"Tonight." He permitted himself a fleeting smile. "You, Mason. Now we'll find out if you've been able to learn anything."

That meant I would be manning a real circuit with operators at the other end, not a passive Fox or weather sked. I had the same sinking feeling in the pit of my stomach that I got when I heard the bugle call for general quarters. I knew that Chief Reeves would not have done this to me, but by now he was getting the radio gang in the *New Mexico* squared away. Still, I wasn't about to help this petty tyrant enjoy himself at my expense.

"That's great!" I said, convincingly I hoped. "Thanks a lot, Schrader."

"Don't mention it, Mason," he said sardonically.

Since I couldn't sleep, I spent the evening kibitzing in main radio and reported to the NPC watch supervisor at 2315. He was glad to see me, explaining that a couple of thirds were in the dispensary with the flu. After giving me a brief tour of the shack and a cup of coffee, he told me I would be on the emergency circuit that linked NPC with all navy ships in the Puget Sound area.

The third class on the all-ships frequency was equally friendly. "Nothing I like better than a man who relieves the watch early," he smiled. "You buckin' for second?"

I didn't tell him it was anxiety rather than ambition that brought me to NPC long before midnight.

"Shouldn't be much traffic on the mid," he said reassuringly as he handed me the headset.

After main radio in the *California*, the NPC radio room seemed very large, with operating positions (mostly unoccupied) against the bulkheads and grouped in islands around the supervisor's desk. It was a quiet watch, and a bull session was in progress most of the night. I didn't dare join in, for fear I would miss a weak call signal. All I could hear in the earphones was the almost inaudible hum of my receiver. I prayed it was tuned correctly. Shore stations used different radio equipment than navy ships, much smaller and lighter, since it didn't have to resist the shock of main-battery fire.

About 0330 the supervisor came over with a message form in his hand. "Let's keep the boys from falling asleep," he said cheerfully. "Send 'em a little test message."

Not knowing I had never before sent a message over a real radio circuit, he left me and rejoined the bull session. I flipped the switch behind the key, and the glowing red light told me the transmitter was live. Taking a deep breath, and grateful I was not under observation, I sent the short message with only a couple of busts. The feedback in my earphones told me it was going out on the air. To my great relief, a series of Rs (for "message received") came back from the *Saratoga, Nevada, Pyro,* and other ships present in the yard. Only a couple of fleet auxiliaries responded with an IMI (a request to repeat certain parts of the message). Feeling quite proud of myself, I signed off my first navy transmission.

When I returned to the ship, I encountered Schrader in the compartment. "How'd it go, Mason?"

"All right," I said happily. "Quiet night. Only sent one message."

He looked disappointed. "Lucked out, didn't you, Mason? I'll see you in emergency radio this afternoon."

I had indeed lucked out. In fact, I had been lucking out ever since I walked into the guard shack at the San Diego Naval Training Station. I should have been pleased. Instead, I kept wondering how much longer my good fortune could continue. It was very easy to get into trouble in a navy battleship. My leading petty officer cordially disliked me, not in a personal way but as a reserve who hadn't earned his rating; he would have been delighted to put me on report for insolence or some other vague charge, especially now that Reeves wasn't around.

For nearly six months I had been keeping my hot temper under iron restraint and acting obedient, respectful, and humble. I wondered how much longer I could maintain what often seemed like a demeaning charade. All I had to do was tell my instructor what I really thought of him, or take a swing at one of the many petty officers who had hazed me, or simply return from liberty a few minutes AOL through not getting a hotel wake-up call. Then I would be mustered at captain's mast like a petty criminal.

The justice system in the navy didn't operate like the one practiced by civilian courts. There, an assumption of innocence was made. Here, the assumption was that you were guilty as charged. Except in rare cases, no allowance was made for mitigating circumstances. You were guilty of being thirty-one minutes AOL, as logged by the JOOW, or you were not. You were guilty of telling your leading petty officer that he was a hostile, surly prick, or you were not. You were guilty of landing a haymaker on the jaw of another member of the naval service, or you were not. The provocation was irrelevant. A prudent sailor manfully admitted his crime and threw himself upon the mercy of his judge and jury, the commanding officer. In the case of our stone-eyed captain, I felt certain the quality of his mercy was highly strained. I prayed, therefore, that I could continue to keep my resentment in check. I did not think I could handle either the stifling confinement or the indignities of the brig.

Ashore, the uninhibited liberty activities of my shipmates seemed pointless and demeaning in their own way. On board, I thought too much about the girl in Placerville and looked too often at her picture, the tiny, five-for-a-quarter photo-booth reproduction of her laughing at me off camera that I carried in my billfold. Now she was nineteen and a wife and mother, but always in my imagination she was exactly as she appeared there: hazel eyes sparkling with promise; teeth untouched by dentist and perfect; long, wheat-blonde hair fussily curled with rollers at the front but oddly becoming; seventeen years old and lovelier than any woman I had ever met, or ever would. The knowledge that we had been too young to cope with the unrelenting opposition of my foster-parents on the one hand, and the pressures of economic survival on the other, did little to lessen my regrets and my depression.

Instead of going ashore, I spent much of my free time with Daley, who was still at the receiving station. Seated on opposing lower bunks, or on the front steps of the barracks if the weather permitted, we spent many pleasant hours drinking coffee and talking with greater enthusiasm than knowledge about the worlds of literature, philosophy, and religion. Daley had made a friend, Sullivan, one of those uncredentialed but perceptive and widely read scholars who could be found in navy enlisted ranks in those days, and he often joined us.

One evening Sullivan quoted a poem that, beginning with a rejection of despair and ending in stout resolution, affected me powerfully.

A *Psalm of Life* should have been familiar to me from high-school English classes, but the only work by Henry Wadsworth Longfellow I could recall was *Song of Hiawatha*. I made Sullivan write the poem down; by the time I left the barracks at tattoo, I had it memorized. I walked back to the ship reciting the lines, particularly the following:

> In the world's broad field of battle,
> In the bivouac of Life,
> Be not like dumb, driven cattle!
> Be a hero in the strife!

That is what I resolved to do, with Longfellow and the Reverend Bayard as my guides. At the hoarse, sadistic shout of the police petty officer, accompanied by the rasping of his baton across the mattress springs,

> All right, you swabbies,
> Reveille!
> Let go a' your cocks
> And grab for your socks!

I sprang from my bunk with my own silent battle cry:

> Let us, then, be up and doing,
> With a heart for any fate;
> Still achieving, still pursuing,
> Learn to labor and to wait.

On board, I labored strenuously at the code machine and practice oscillator to become a more proficient radio operator. Ashore, there were wholesome, healthful activities to be pursued while I waited to meet the woman who would replace the one of the lost dream.

I worked out with the *California* swimming team in the basement pool at the YMCA and easily qualified for the fifty-yard free style. It wasn't that I was that good—only dedication and practice would have elevated me to tournament class—but that the ship's team suffered from a shortage of talent. One notable exception was our interim coach, boatswain's mate second class L. P. ("Speed") Spadone, who swam the 220- and 440-yard free-style events and the relay.

Occasionally, I sparred with members of various ships' boxing teams that trained at Craven Center, dedicated and clean-living fellows who were really good. I was a better-than-average boxer but, as with swimming, hadn't dedicated myself to it. Consequently, I was just proficient enough to serve admirably as a moving target. (These exercises in masochism had a double advantage: they conditioned one physically and helped one atone for past sins.) Although team athletics was being deemphasized in the fleet in early 1941, boxing was still a major sport, and many navy boxers were easily the equal of professionals.

Observing the special treatment the boxing team received in the *California*, sophisticates like A. Q. Beal—a burly, poetry-writing second-class radioman with a college background and one of the highest I.Q.s in the radio gang—commented that our boxers were a long step removed from pure amateurs. They were excused from many watches and other duties so they could work out every afternoon at 1400 in a well-ventilated second-deck compartment. They had their own mess, where they enjoyed fresh

Action was seldom lacking at navy boxing smokers. This one was probably held at Craven Recreation Center, Puget Sound Navy Yard, circa 1941. The sailor bringing up a sweeping left hook has "Tippecanoe" and "Donna" tattooed on his left shoulder and arm. Boxing was one of the few activities that was open to all ratings in the almost totally segregated navy of the author's time. (Photo courtesy of the Bremerton Armed Forces YMCA.)

milk and eggs and tender steaks in place of the powdered stuff and the tough, dry "ox meat" that was served in the crew's mess in Hawaiian waters.

The head coach of the boxing team was John E. ("Spud") Murphy, an Old Navy chief boatswain's mate who had served under General J. J. ("Blackjack") Pershing in the punitive expedition against Mexico's Pancho Villa in 1916–17. Murphy was so tough, according to the legend, that he chewed up stanchions and spat out iron filings. He had been sent to the *California* as chief master-at-arms under specific orders to clean up the ship's petty loansharking, gambling, and laundry and liquor rackets, which had proliferated under lax commanding officers during the lean days of the Great Depression.

As the number one enlisted man in the ship, Murphy lived in lordly splendor in his own stateroom, near the bow on the main deck. He was assisted in his coaching duties by "Biff" Bailey, BM1, a former all-navy light-heavyweight champion, and "Peg" Brown, MM1, as well as a trainer who also had the ship's newspaper and magazine concession and was known to all as Sweatshirt. The fact that the boxers operated under the aegis of the ship's "top cop" added considerably to their privileged status.

But I had more important things than the pugilists to consider. While they followed their high-protein diets and abstained from most, if not all, vices, I was seeking nourishment of the spiritual and cultural kinds.

I attended Christian Science services on Sunday mornings and accepted invitations to liquorless dinners in the suburbs. The good Bremerton burghers were more tolerant than those in San Diego had been: I was served coffee, and ash trays were brought out and dusted off for my use.

Boxing team of the *California*, 1939–40, photographed on the ship's forecastle. Fifth from the right is Vernon L. ("Lute") Luettinger, RM3, BatDiv 2 bantam-weight champion. At the far right stands head coach John E. ("Spud") Murphy, a former boxer and the ship's feared and respected chief master-at-arms. (Photo courtesy of Vernon L. Luettinger.)

One evening I went to Seattle's plush Fifth Avenue Theater, where I saw the famous actress Tallulah Bankhead, live and in full throat, in *The Little Foxes*. I sat in the seventy-five-cent seats, in the highest level of what was then called nigger heaven, marveling at her diction, every syllable of which was perfectly intelligible, marveling even more that Miss Bankhead was getting by on personality and her unforgettable throaty voice, not on talent.

As I left the theater, I suddenly realized I felt fine. The murky depression through which I had been groping the past month had lifted like the morning overcast at San Diego. Whether Miss Bankhead deserved any of the credit, I didn't know. I thanked her lobby photo, just in case, and went looking for my friends.

Later, I understood that I had been at a critical point in my adjustment to the battleship navy, that point at which many others, including some currently in the brig, in the hospital, or awaiting discharge under conditions other than honorable, had failed. My casual friend Sullivan at the receiving station had given me, at precisely the right moment, the key to triumphing over the regimentation while still bowing to it. I did not have to permit the system to break my spirit and reduce me to a rate and service number, nor did I have to become a "rebel without a cause." Having found that middle ground, it was possible to enjoy some of the best that navy life had to offer, always maintaining a reserve of fortitude that could be called up to withstand the worst.

I sat with a couple of friends in a dim basement beer tavern on Second Avenue and celebrated my return from voluntary exile with a Coke. The place was full of sailors in dress blues and the rakish flat hats that still, in early 1941, bore the ships' names on the ribbons.

Mingling with the sailors were a goodly number of sailors' girls. These casualties of the Great Depression were drawn mostly from the lower middle class. They came out of high school—if they attended secondary school at all—with no training, no job skills, and no opportunities. In navy towns like Seattle, many of them drifted into brief liaisons with sailors they met in bars. A few of the more fortunate ones found enlisted men who would take care of them, sometimes with allotments and sometimes even with marriage.

Here, among the unattached, one looked almost in vain for beauty. Incarnate were all the evils of America's continuous neglect of its poor, especially during the past depression decade: bad teeth, poor carriages, lusterless hair, bowed legs, faces puffy with starch or gaunt from early malnutrition. Only occasionlly would one see an attractive girl, and she, invariably, would be surrounded by uniformed Lochinvars. Not that the others lacked for attention. Here they were all Helens of Troy, and scarcely one would fail to find a sailor eager to make love to her. Often more than one sailor was willing, and I would be treated to the extraordinary sight of two handsome and virile enlisted men fighting over a woman neither would have favored with a second glance in his home town.

Months before Johnson told me that in a lousy navy town like Honolulu you had to make do with what was available, I decided the navy knew a hell of a lot about running a military organized along the lines of a caste system. First, you branded a man with inferiority based upon his family background, his education, and the type of uniform he wore. Then you reinforced it with every means at your command, including cramped living quarters, total lack of privacy, and continuous musters and drills, moving up to the report, the captain's mast, the brig—and the short-arm inspection. (As a means of impressing a man with his inferiority, nothing exceeded the humiliation of the short-arm inspection en masse, as I shall recount.) After a man had begun to accept his inferiority, you sent him ashore with carte blanche to indulge it. Before long, *voilà*! He not only accepted it—he gloried in it.

To prove my thesis, all I had to do was look down the bar at my high-school buddy Fisher, aggressively pawing a waterfront floozy. Although his talents would later be recognized when the navy sent him to officer candidate school at the famous California Institute of Technology in Pasadena, he was, at this moment, thoroughly enjoying his inferiority.

It was obvious that I would never begin to enjoy mine unless I stopped drinking Cokes, going to the Fifth Avenue Theater, and acting like a refugee from a small-town DeMolay chapter (which, in fact, I was). I ordered a bottle of beer, discreetly, I thought, but the word was quickly passed. Suddenly, I was surrounded by grinning shipmates, slapping me on the back, mussing my hair, and telling me it was about time, by God, that I joined the real radio gang, and the real navy. Basking in this camaraderie, I drank more beer than a hero of the world's broad field of battle should, and compounded the sin by switching to whiskey when our large and raucous party adjourned to the Blackstone Hotel. The next day, I couldn't have given a full account of the evening without the help of my buddies. But certain events remained clearly in focus. I had proved my conjectures about the caste system. Inferiority does have its own rewards.

Just as it has its hazards. I never did find out who stole my flat hat.

Not long afterward I monitored an upper-division course in the education of an enlisted man. The class was held after hours in a bottle club called The Breakers, somewhere in Old Town.

It was a grim, dank, hotly lighted room, barren of furnishings except for heavy, cafeteria-style tables and benches—a most practical decor, as events soon proved. While it served food during regular hours, its *raison d'être* was thirsty sailors after the legitimate beer taverns and dance halls of

Seattle had closed. You brought your own bottle, purchased from one of the state-operated liquor stores; the proprietors sold setups of ice and mixer at something like fifty cents a glass. A large juke box gobbled up the sailors' nickels and dimes; a number of navy girls added a certain spice to the atmosphere.

My tutor was Johnson, now released from twenty days in the brig and at liberty from his steady girl friend Eleanor for a few hours. When the first setups arrived, he took a pint of bourbon from a pocket of his peacoat and poured generously into our glasses.

"Since you haven't been here before," Johnny said after the first skoal, "here's a little advice. When the trouble starts, follow me under the table."

"What kind of trouble?" I asked naively.

"Look around," he replied. "What do you see? A hell of a lot of sailors—and not too many broads. That always spells trouble. When the bottles start flying, we take cover."

"They throw bottles?" I asked incredulously.

"Damn right. First, the swab jockeys start swinging. That leads to ash trays and bottles. Me, I give 'em plenty of gangway. I never fight unless it's really important." He laughed, took a long drink. "You know me, buddy. I'm a lover, not a fighter."

Johnson's reputation in that area was well established. I wondered what he thought truly important, and asked him.

"Only three things, Ted," he said, serious for a moment. "Defending your country. Defending your woman. Looking out for a buddy, if he needs it. All the rest is bullshit." He poured more bourbon. "Now, if you've got a good woman, like Eleanor, she won't get you into that kind of trouble. So you're down to two. Only two, my friend."

But others at The Breakers did not share Johnson's viewpoint. The trouble began, predictably, with loud voices from a long table occupied by too many sailors and too few girls. The voices became louder and more profane, and there was sudden movement. A boatswain's mate stood up and aimed a big, looping punch across the table. His momentum carried him right along and he tumbled against his adversary. As bottles and glasses spilled and the women fled, the two fell out of sight, pummeling each other.

Johnson watched all this with a half-amused, half-disgusted smile. "Stand by, Ted," he advised, hastily draining his glass and stowing the bottle in his peacoat. "There goes the evening."

In a trice, the action had entangled the whole table. Inevitably, the territory of the combatants spread and involved the other tables, in a simple demonstration of great-power politics. The prudent and the neutrals retreated from a dispute that was none of their concern. The hotheads and those with partisan loyalties thrust themselves or were drawn into the fray. As the action became general, the absence of the hysterics that usually accompanied civilian melees seduced me into believing I could remain a mere observer at the finest pier-six brawl I had ever seen. The men fought not only without histrionics but also with the kind of fierce joy to be expected of those who had been trained for combat but long denied that ultimate demonstration of their professionalism.

I was seconds away from being engulfed by the tide of swinging, ducking, surging, retreating, falling sailors when Johnson grabbed me and pulled me down.

"Take cover, God damn it!" he cried. Sure enough, we had no sooner settled under the solid oak table when a whiskey bottle bounced off the topside and shattered into a hundred flying fragments.

As fists flew in the center and glass from the periphery of the moving front, the doors burst open and the peacemakers stormed in, shouting, "Shore patrol! Break it up, goddamit, break it up!"

With remarkable alacrity, the combatants disengaged, a few in mid-swing. Breathing hard, they slipped back to their tables with skinned knuckles, bloody mouths and noses, and assorted other wounds, some with embarrassed smirks of satisfaction, others in scowling and wordy disgruntlement, depending upon how they had fared. Most of the SPs left before long, triumphantly escorting a few battlers who had had nothing to do with starting the fight but lacked the good sense to know when to stop. The real perpetrators melted into the crowd and escaped detection.

Johnson and I emerged from our improvised air-raid shelter and poured ourselves a soothing nightcap while the manager and a couple of SPs who had remained behind tried to clear The Breakers. My friend began to laugh.

"Look," he said. "The broad who started all this is leaving with another sailor." An attractive blonde, who had been seated next to the boatswain's mate before he followed his fist across the table, was moving unobtrusively out the door with a lean, dark-haired young petty officer. He had been nowhere near the scene of action.

"Serves them right," Johnson said scornfully. "Fighting over a little chippie like that. Don't they know Seattle is full of them?"

February passed into March and we remained moored port side to Pier 5, Berth D. During this time, we got a new executive officer. Commander Carney was ordered to Washington for duty with the Chief of Naval Operations. Arriving from a CNO staff job three weeks later was Commander Earl E. Stone. We shrugged. Under a skipper like Bunkley, what difference did it make?

Nine months later, at Pearl Harbor, we found out what a difference one officer can make—what a difference, in fact, one chief quartermaster can

make. Learning that, I wondered whether the *California* would have fared
better had Carney still been exec. Would he have been on board to take
command during that critical first hour, as Stone was not? Or would he,
too, have acquired a permanent blot on his record, a blot that would have
prevented him from ever serving as Halsey's Chief of Staff or as CNO? To
reach the heights, a navy officer must not only be good, he must also be
lucky. Carney was both.

In mid-March a reserve officer reported for duty from the Twelfth Naval
District in San Francisco. Although he was only a lieutenant junior grade,
his arrival was a matter of great interest to the radio gang. For Proctor A.
Sugg was our new radio officer. Our previous one had been so bland and
self-effacing that I can't even remember what he looked like, let alone his
name. But Lieutenant Sugg soon made his presence known.

He was a tall, well-proportioned blond man in his early thirties, with
near movie-star good looks. His uniforms were tailored, his grooming was
always immaculate, and he glowed with health and energy. He was no
typical ninety-day wonder, the scuttlebutt informed us: he had been a
big-shot executive with NBC radio network in San Francisco. Looking up
from my typewriter, I would see Sugg circulating in the comm office and

Author's radio officer, Proctor A. Sugg (*right*), with unidentified lieutenant commander. Photo was taken in late 1944 or early 1945, when Sugg was commanding
officer at Ward Island, Corpus Christi. He left the navy as a captain, USNR, and was
executive vice-president of NBC Television from 1958 until he retired in 1962. He
died in 1976. (Photo courtesy of Mrs. Betty Sugg.)

main radio. He introduced himself all around, learned each man's name and something of his background—and did it without condescension. I was impressed, as were most of the radiomen. If one needed a model for an officer and a gentleman, I thought, Lieutenant Sugg was the man.

Toward the end of March, the rising tempo of events heralded our imminent departure from Bremerton. A party of forty-two boots reported in from Great Lakes Naval Training Station. A half-dozen newly hatched ensigns, all USNR, came aboard from the naval midshipman school in Chicago. Our navigation officer, first lieutenant and damage-control officer, and medical officer were replaced. And some of our enlisted foul balls were weeded out.

Two seaman seconds and a mess attendant third class were sent to the receiving station under armed guard, two to await bad conduct discharges and the third a general court-martial. The unfortunate marine private who discovered that even illness could not shorten his brig sentence had once again failed to find his middle ground and was escorted to the marine barracks for his bad conduct discharge.

For half a year, Fisher, Stammerjohn, and I had been carrying the tattered banner of the V-3 Communication Reserve. On the last day of March, we received welcome reinforcements. From the naval reserve training school at Noroton Heights, Connecticut, by way of the receiving station at San Diego and the USS *Neches* (AO 5), came third-class radiomen David ("Rosey") Rosenberg, Julian (Jack) Rossnick, and A. A. (Al) Rosenthal.

In early 1941, there were few Jews in the navy, and practically none in the battleship navy. Here were three of them in one radio gang, all with some college and engineering backgrounds and all well grounded in radio code and theory. Into the predominantly Anglo-Saxon C-D Division they brought the accents of Brooklyn and the Bronx, harsh-sounding and nasal to the ears of a Californian, a certain separateness of culture, and an individuality of thought and action that made their adjustment to life in the *California* at least as difficult as mine had been. While they lacked my considerable advantage of a name short, simple, and unquestionably Anglo-Saxon, they already possessed the radio operating and repair skills I was still struggling to acquire. And, just as I had, they benefited from fair and just treatment by Chief Reeves.

It was probably only because of the Great Depression that Rosenberg and Rossnick were in navy enlisted ranks. Both nearing their late twenties, they were among the countless urban casualties who had seen their hopes and ambitions for professional careers blasted by lack of educational funds and employment opportunities.

Rosenberg was introverted, tidy, and rather critical by nature. His principal complaints about the *California* were having to stand in line for

"The *California* Issuing Room sizing up well stocked shelves" is the typically optimistic *Our Navy* caption for this 1939 photo. In 1941, the author's shipmate Jack Rossnick described the chow with considerably more realism as "ground-meat sauce on toast, canned horsemeat, roasted vulture and dishwater soup." (Photo courtesy of the Naval Institute Collection.)

all necessities and being chased from compartment to compartment in his quest for a private place to relax. Rossnick, full faced and of rather pudgy build, was more cheerful and outgoing. But he was cursed with a nervous stomach that rebelled against the standard shipboard chow, which he vividly described as "ground-meat sauce on toast, canned horsemeat, roasted vulture, and dishwater soup."

The extroverted, high-spirited one of the trio was 21-year-old Al Rosenthal. He was nearly six feet tall, handsome, and olive-skinned, with masses of black, curly hair. Unlike the conscientious, hard-working, non-drinking Rosenberg and Rossnick, he was careless and somewhat irresponsible in the performance of his duties, preferring to get by (not always successfully) on his ingratiating personality. On liberty, he devoted himself to drinking and seduction, two activities that gave him common ground with a number of his fellow radiomen.

As we made all preparations for getting under way, $4.5 million in cash was brought on board for transfer to the fleet paymaster upon arrival in Pearl Harbor. All who were topside stopped whatever they were doing to gawk as the money arrived with an escort squad of our biggest, toughest-looking marines. It was a great deal more money than any enlisted man had ever seen, and the scuttlebutt raced from the quarterdeck to the double bottoms.

"Jesus Christ! We just took ten mill aboard! In cash!"

By the time the word had been passed up to the foretop, the amount had increased to $20 million. If our President had just been piped over the side, the crew could hardly have been more awed and respectful. There was much discussion of what one could do with even a small percentage of that much "loot."

"God damn!" one radioman intoned reverently. "That would be even better than marrying a wealthy nympho who owns a bar and liquor store!"

There was one slight drawback to having $4.5 million in the ship's safe under marine guard night and day. It meant we were returning to grim, gray Pearl Harbor, where Kimmel had taken over from Richardson, and where Snyder had been relieved by Vice Admiral W.S. Pye, who had fleeted up from ComBatShips.

I consoled myself with another choice morsel of scuttlebutt. We were stopping in San Francisco. If I continued to be lucky, I would see the Golden Gate from the sea.

CHAPTER 8

# Return to Pearl: "This Bloody Awful Hole"

Unlike most wonders of nature, the Golden Gate has actually been improved by man. Coming in from the Farallon Islands, there seems to be no break in the continuous line of Coast Range foothills. Then, hazy and shimmering like a mirage, two graceful thin towers and a low connecting deck emerge—a 4,200-foot-long passage to nowhere. As the ship skirts Four Fathom Bank and enters the narrow channel, visibility improves. The Marin Peninsula to the left comes tumbling down from Mount Tamalpais and disappears into the ocean at Point Bonita. On the right, the jagged spires of Mile Rocks warn of Land's End. Into the break between flows the now-turbulent Pacific. Linking the two promontories at their closest approach is the Golden Gate Bridge, simple and clean and soaring, a fitting frame for the strait that opens into one of the world's largest and finest natural harbors.

I stood on the boat deck as the *California* approached the bridge, aiming her clipper bow at the three vertical white lights in its center, where the suspension cables swooped down to meet the gently cambered roadway, and knew this was one of those moments I would always remember. The news of our arrival must have preceded us, for a good number of spectators had gathered on the pedestrian lanes of the bridge. As we passed beneath them, our topmasts appearing to clear by only a few feet, they merrily waved arms and hats and a few flags. I could see that even the cynics among us were stirred. We waved back, some a little embarrassed at such a public display of emotion, and thought of liberty in this fabulous city of hard-drinking men, extraordinary food, and beautiful, willing women, just as sailors had always done.

We passed the cool, landscaped terraces of the Presidio Military Reservation, so inviting that one almost wished to be in the army (almost but not quite), and now changed course to pass well clear of Alcatraz, rising up in the middle of the bay as one approached from the sea.

Stark white buildings protruded from the tiny island like quartz out-croppings in a gray rock. And The Rock was in fact its nickname. Thanks to the efficient public-relations arm of J. Edgar Hoover's FBI, every good citizen was aware that self-effacing but implacable agents, under the inspired leadership of their director, had rounded up the nation's top criminals and confined them to this American Devil's Island, where by a cruelty perhaps unintentional, they could gaze daily on America's fairest city and rue the penalties of crime.

Everyone on the boat deck was staring at Alcatraz. "Jeez!" one seaman exclaimed. "The Rock. That's where all the tough guys are. Al Capone. Machine Gun Kelley. . . ."

A first-class boatswain's mate, standing nearby, overheard him. "Tough, hell!" he said contemptuously. "They're nothin' but punks. Take away their guns and their goons and they wouldn't last a week in the battleship navy."

The boats, six feet of rock-hard muscle, was a former fleet wrestling champion who bossed a deck division. He was tough in the best sense of the word. No one disagreed with him.

As we passed the fishing boats tied up at Fisherman's Wharf, the full panorama of San Francisco came into view. A little to the southeast was Telegraph Hill (where in gold-rush days a semaphore station announced the arrival of the ships and delivered such useful information as how many women were aboard), surmounted by the fluted white cylinder of Coit Tower. To the west of this solitary eminence, the city's central district rose tier after tier above the bay, looking like the world's largest and whitest wedding cake.

As the ship changed courses with the channel, amoeba-shaped Yerba Buena Island appeared off the port bow. Everyone with any navy service knew about "Goat Island": a tumbledown wooden receiving station was located there. The building dated back to the Spanish-American War, when it had been constructed as a naval training station. In 1923 the training station was relocated to San Diego, where there are no damp cold fogs and cruel punishing winds like the ones on Yerba Buena that had put half of every recruit company into sick bay. (While I could not know it then, that primitive barracks would be my home when I was assigned to new construction in the summer of 1942. Shore-patrol duty in San Francisco came as a most welcome assignment.)

Moored to a nearby pier was an even more venerable relic of the "American Steel Navy": the protected cruiser *Boston*, built in the 1880s, a veteran of Dewey's squadron at Manila in 1898 and now classified as a "receiving ship." On a hill above the base, the commandant lived in a splendid green and white Victorian mansion.

Attached to Yerba Buena by a causeway was the world's largest man-made island, a seven-sided geometric figure named Treasure by someone with an impish sense of humor. It had been the site of the 1939 Golden Gate International Exposition, which I had attended with Fisher and his family. After the exposition ended, the island had no useful civilian purpose, being plagued by the same fogs and winds that had driven the training station off Yerba Buena. The navy took it over. (I was later to know "Treasure Island" rather better than I might have wished, had I not been aware of the alternative: meeting with the kamikazes off Okinawa.)

Yerba Buena played another role: it anchored the two long sections of the San Francisco–Oakland Bay Bridge, now looming before me. As a sheer engineering project, the bay bridge was even more impressive than the Golden Gate span: it reached four and one-half miles over water, with every foot a visual delight of form as perfectly suited to function as a *Fletcher*-class destroyer.

We passed under the bridge and anchored off Rincon Point. The only other navy ship present was the *King* (DD 242), a 1920 flushdecker, and the *California* crew could barely contain its excitement at the prospect of liberty in a city that was not swarming with sailors. Alas for the best-laid plans of mice and enlisted men: the word was passed that we were leaving San Francisco the next day, and only one watch (about half the crew) would be granted liberty.

My luck had deserted me: I was in the wrong watch. For a long time I stared at the skyline of "The City." Market Street made a thin diagonal from the graceful Moorish Ferry Building at its foot to its termination near Twin Peaks, which rose brown and treeless in the center of the peninsula. On the near flank of the peaks, in a typically tall, thin San Francisco house, its gray clapboard facade embellished with gingerbread, lived my father and his second wife, Carol, and my half-brother and -sister, Richard and Joan. I would not see them this time.

We took our departure for Pearl Harbor early the next afternoon. I was at the code machine when our philosopher second class burst into emergency radio.

"Jesus Christ!" He was almost shouting in his excitement. "We damn near got us one of the President liners! Passed so close aboard I could practically reach out and shake hands with the passengers!"

"How could that happen?" I asked, shutting off the code machine.

"Easy on the Prune Barge! Our goddam officers have got us steaming down the wrong side of the channel! Passed a freighter and a small coastal liner and the big one starboard to starboard! Jesus Christ!"

If the *California* was passing starboard to starboard, we were in gross violation of the rules of the road. "Are you sure?" I asked.

"Damn right I'm sure! Didn't I see it with my own eyes?" He paused to catch his breath. "Now BuNav has sent us a navigator who can't read a chart! Soon as we reach Pearl—if we reach Pearl—I'm gonna put in for a transfer. Jesus Christ!"

The *California* steams under the Golden Gate Bridge, inbound for what was then one of the world's great liberty ports. (Photo courtesy of the Bear Photo Service.)

Although convinced the *California* was on a collision course with disaster, our excitable radioman never did ask for a transfer. Rather, he seemed as fascinated by the prospect as a bird hypnotized by a coiled snake. Or perhaps he couldn't resist staying around to say to the survivors, "I told you so!"

Shortly after we got under way, the scuttlebutt ran through the ship that a civilian had come aboard just before we weighed anchor and been quickly escorted to the wardroom by a phalanx of officers. A civilian in a navy ship was such a rarity in the spring of 1941 that he was obviously a big shot, possibly from the secretary of the navy's office. An observant sailor pointed out, however, that no four-star flag with the fouled anchor on a blue, red, or white field was flying from the main, so he couldn't be SecNav or one of his two top subordinates, the undersecretary or assistant secretary. Soon the word was passed along from the strikers who ran the decoded messages to officers' country: our guest was Walter Davenport, the famous writer from *Collier's* magazine.

*Our Navy* was the publication most of the crew read, if they read anything at all beyond comic books or the correspondence courses for advancement in rating. But *Collier's* was known to all of us, since it ranked with the *Saturday Evening Post* as one of the two most popular and prestigious mass-circulation magazines in the country (the newsstand price of each was five cents). Its oversized, slick format combined fiction, articles on current events and personalities, cartoons, and humorous poetry with a great many advertisements for automobiles, cigarettes, toothpaste, and soap. Its glossy, process-color covers decorated the coffee tables of millions of the nation's homes and offices—and Walter Davenport was its star feature writer.

For a couple of days I kept a sharp lookout topside for this big-time journalist from glamorous New York. If he had been John Steinbeck or Ernest Hemingway, I was sure he would have wasted no time escaping the stuffy wardroom so he could mingle with the men who truly represented America: the crew. There was, of course, another possibility. Writers reputedly drank even more than sailors. Moreover, they were immune from watches, musters, drills, and captain's masts. Davenport could be holed up in some third-deck stateroom, gloriously drunk and tapping away at his Great American Novel.

By the third evening, the suspense was unbearable. Putting on a clean, pressed uniform of undress blues and a neckerchief, I went down to radio one and "pulled rate" on the messenger of the watch. The young striker was only too happy to turn over the duty belt, leggings, and message clipboard that would give me entree to otherwise restricted territory.

The decks in officers' country were covered with thick, highly polished battleship linoleum in familiar rust-red. Bulkheads and overheads were painted a contrasting pale green. The metal doors to the staterooms were kept discreetly closed, and the hushed atmosphere seemed charged with tension. Treading these passageways as an enlisted man, one felt small and insignificant, a spearholder in the councils of the mighty.

My first stop was the captain's cabin. Naturally, messengers were never permitted inside this sanctum sanctorum: the skipper's marine orderly took the messages the last few feet. This one was typically tall and lean, chewing gum with an occasional slow twitch of a lantern jaw. He was standing at parade rest in an immaculate full-dress uniform, doing his best to look forbidding.

"Watcha got there?" he demanded in a hard cop's voice.

I decided I was not going to allow a mere marine private to intimidate me. Besides, I had a mission: to find the great Davenport. "Message for the captain," I said shortly.

"Gimme it."

I started to detach the top message from the clipboard. "The whole goddam thing, sailor. The captain likes to know what's goin' on around his ship."

I handed him the board, and he stared at it in the myopic fashion of people who do not read well.

"Want to borrow my glasses?" I asked.

He gave me the marine glare second class. "It don't look very all-fired important to me," he said. "The skipper may not want to be bothered tonight."

"He hungover?" Captains, from what I had heard, drank a lot too.

The private shot me the first-class glare this time. "That's how much you know," he said, snapping off each word as if he were counting cadence. "It so happens he can't drink."

"Yeah?" I didn't have to try to look surprised.

He tapped his stomach. "Bad gut. He's been under the weather since we left the States."

I was beginning to get some good scuttlebutt. "Ulcer?"

"What the hell you think it is—a hernia?"

"Me, I thought he was just a natural S.O.B. I didn't know he had an ulcer."

"You swabbies don't know nothin'. With all the eightballs like you on this ship, he's gotta be an S.O.B." He paused, then started up on another tack. "Hey, I bet you don't even know the old man wrote a book."

Again, I was surprised. "He wrote a book? What is it, a biography of Captain Bligh?"

That went over the orderly's head. "Naw, it's called the *Naval and Military Recog*—uh, *Recognition Book*—somethin' like that. There's a copy in the library. You ever visit the library, Mac?"

"Hell, no. Too many sky pilots around there." I decided to try for more inside dope. "Say, I'm sorry to hear about the skipper. Maybe they'll give him a medical."

I don't think I was able to keep the hope out of my voice. The marine's already narrow eyes became mere slits. Obviously, there would be no more scuttlebutt.

"What's it to you? You plannin' to take over the conn?" His shoulders shook with soundless laughter.

"Yeah. And you know the first thing I'm gonna do? I'm gonna send the whole frigging marine division to Wake Island." I paused for dramatic effect. "Then I'm gonna replace 'em with a detachment of hospital corpsmen."

Now we were eyeball to eyeball, and his were bulging. Without a word he about-faced, knocked on the cabin door, paused for permission, and entered. When he returned, he thrust the clipboard at me as if it were a bayonet.

"I'm gonna remember you, Swabbie," he snarled, eyeing my chevron. "What's your name, rate, and service number?"

"What's it to you, private?" I riposted. "You gonna try to put a third-class petty officer on report?" I swaggered aft, leaving him choking in his own bile.

My next stop was the wardroom, where I had been told I would find Lieutenant Sugg. In the *California*, which had been designed as a force flagship, the wardroom was large and luxurious. It ran clear athwartships on the second deck, abaft number four turret barbette. The atmosphere, I judged from my readings in Somerset Maugham, was that of an exclusive men's club.

On the starboard side were sofas and easy chairs in soft, rich leather. The ports were draped with heavy dark-green curtains. On the other side were refectory tables covered with green baize and decorated with silver bowls piled high with fruit. Filipino and Negro steward's mates in starched white uniforms stood by in respectful attitudes, ready to hurry over with refills from silver coffee pots fitted with long wooden pouring handles.

Seated around one end of the longest table were Sugg, several other officers, and a lone civilian in a gray business suit, who had to be Davenport. His face was lined and sallow; his hair was nearly white. He seemed incredibly old to my young eyes—even older than the skipper—and not at all as I had imagined a famous writer would look.

Being in enemy territory, I approached with trepidation. Davenport was telling a story, and the officers were leaning forward eagerly, chuckling in appreciation as if he were a captain or an admiral. Sugg spotted me and waved me forward. Using my full name, he introduced me to the great man as one of his bright young radio operators. The writer flicked his eyes in my direction.

"Hi, Ted," he said in an offhand way, and immediately resumed his anecdote about some prominent political figure.

As soon as Sugg signed off his messages, I escaped the wardroom. I was disappointed by the outcome of my meeting with Davenport and decided there was something essentially phony about him. He didn't look like a real writer—and he didn't seem to act like one, either. He's sure as hell no Steinbeck, I told myself. He's no Hemingway. And he'll never write any Great American Novels.

What Davenport did eventually write was a magazine article for the June 14 issue of *Collier's* entitled "Impregnable Pearl Harbor." I read it with great interest and increasing astonishment at the journalistic distortion that seemed to pervade Davenport's prose.

The subhead at the top of the article claimed he had "talked with three admirals" as well as "boatloads of bosuns and their mates." The admirals, one could assume, were Kimmel, Pye, and Claude C. Bloch, commandant of the Fourteenth Naval District. The following excerpt from the article was apparently based on what they told him:

> Day and night, Navy and Army planes are droning down the warm skies in circles two hundred, five hundred, a thousand miles wide. They're dropping bombs from altitudes of twenty and thirty thousand feet and smashing tiny targets towed by swift destroyers. They can't see the targets, half the size of a life raft, but they're sinking them from aloft. And they're blasting them in dives. By radio they're hooked up with infantry, artillery, the Fleet, their brother fliers. The defense of Hawaii may not be impregnable. Ships can be sunk. Planes can be downed. Forts—even Diamond Head—may be razed. And serried ranks of big guns can be silenced. But neither the Army nor the Navy believes that there is any power or combination of powers existing today that can prove it in the Islands.

The mischief of Davenport's article, which was written with the obvious blessing of the navy, was that it painted a false picture of the state of readiness at Pearl Harbor. As all the world knows, there was no long-range aerial reconnaissance worthy of the name. Had there been, the Japanese attack force undoubtedly would have been detected—or, an even more interesting possibility, might never have sailed from the Kurile Islands. When war came, the army bombers that had been smashing those tiny targets at 30,000 feet proved inept against real, maneuvering ships. The radio-communication hookup among the services existed mostly on paper. The operational word in the navy, army, and air corps in 1941 was competition, not cooperation.

The only ones misled by this article were the American people. At faraway Kure naval base on the Inland Sea of Japan, a certain poker-playing admiral who flew his flag from the battleship *Nagato* was not misled.

His name was Isoruku Yamamoto.

Except for some construction activity in the yard and on Ford Island, Pearl Harbor looked much the same as when I had left it more than five months before.

"The bastard who named this cesspool *Pearl* Harbor should be keelhauled!" one sailor groused as he flipped a cigarette into water the color of scorched split-pea soup.

Our customary quay was vacant, but we ignored it and eased alongside Ten-Ten Dock instead, usurping the absent *Pennsylvania*'s berth—the better to discharge our navy passengers, Davenport, and the $4.5 million in cash, I surmised.

It didn't take long to discover that the atmosphere at Pearl Harbor had changed under Admiral Kimmel, the new fleet commander. Without warning, the air-raid sirens sounded on the evening of 20 April, and all the lights of the harbor winked out. In Honolulu, the SPs and MPs chased all the sailors and soldiers back to their ships and bases. Soon Honolulu was blacked out, too. Disgruntled sailors returning to the ship were loud in their condemnation of Kimmel and Bloch for pulling a "boot-camp drill." The most vociferous were those who had been rousted from Hotel, River, and Beretania Street houses of joy, some of them caught, quite literally, with their pants down.

After a week in port, we spent a week at sea conducting various exercises with Task Force Two: four battleships screened by assorted cruisers and destroyers. The only excitement in the routine of watches, drills, and general quarters was the day our VOS plane 2-0-8 crashed on landing in a choppy sea and promptly capsized. Two motor whaleboats were quickly hoisted out, and the pilot and his observer were rescued with no more damage than wet uniforms and injured pride. Another twenty minutes were required to hook onto and hoist in the damaged plane. I was glad to observe that the welfare of the men took precedence over the machine, a navy priority verified many times later in the South and Central Pacific.

When we returned to Pearl on 5 May, the navy yard pilot put us expertly alongside the *New Mexico*, moored to our familiar Berth F-3. Gloomily, we looked up at the three-star flag hanging limply from the battleship's main. We knew it would very soon fly from ours. When the flag radiomen came trooping into compartment B-511-L, we greeted them with ambivalent feelings and did not offer to return their choice lockers and bunks.

Nine days later, we went back to sea for tactical exercises with BatDivs 1 and 2, less the *Pennsylvania*. After five days, we were joined by Kimmel, flying his four-star flag in the *Pennsylvania*, and by BatDivs 3 (the *New Mexico, Idaho,* and *Mississippi*) and 4 (the *West Virginia, Colorado,* and *Maryland*) for "fleet tactics." Now all twelve battleships of the Pacific Fleet were crisscrossing the waters off Hawaii in involved maneuvers and simulated battle-line engagements whose evolutions remained a mystery to me, confined as I was to the third deck and the first platform. I got the vague impression, though, that we were refighting the battle of Jutland.

When I tried to find out, I met with blank hostility. As usual, the battleship navy was operating on the "need-to-know" basis. Since there was little I needed to know to copy NPM Fox, I was told nothing. The prevailing attitude was, What business is it of yours, sailor?

After continual rebuffs of this kind, most crewmen stopped trying to learn anything. They went about their narrowly specialized duties in a state of real or pretended indifference. "The sooner we get this shit over with," they complained, "the sooner we'll return to port."

It seemed to me that we would take more interest in our jobs and perform them with more zeal and efficiency if we knew how they fitted into the overall plans and objectives.

I longed to say to one of my officers: "I'm an American. My family has fought in every one of our wars since the Revolution. We're on the same side, dammit!"

But I had no wish to pay a social call on Bunkley at captain's mast. I contented myself with slipping topside at every opportunity, where I could admire our forces steaming in line of battle ahead, turrets trained out toward the "enemy." The eight to twelve big guns of each battleship could deliver a devastating broadside of about eight tons of armor-piercing steel and high explosive. Hull down on the horizon, the opposing battleships were easily recognizable as American by their tall, symmetrical twin battle tops. This was the way combat at sea had been waged since the seventeenth century. Only the range at which the forces engaged each other had steadily increased, from point-blank to 10,000, then 20,000 and 30,000 yards. Within a year, the maximum combat range would take a quantum leap, courtesy of the aircraft carrier. But we had no inkling of that. The battleship was still queen of the seas: who would dare challenge us?

Our only casualties were two seaman mess cooks. One cut his right hand in a losing joust with broken crockery, and the other lacerated his fingers in underwater combat with a carving knife in the scullery of the CPO mess. Both were treated at sick bay and quickly returned to duty. Another fallen warrior was a torpedoman third class from one of our accompanying destroyers, the *Wilson* (DD 408), who had suffered an attack of acute appendicitis. We stopped all engines, backed down, and lowered the number two motor launch. In a few minutes the boat had picked up the sailor and was hoisted back in as we got under way again. The patient was rushed to sick bay, where he was operated on immediately (and success-fully) in the ship's small but well-equipped surgery.

When we returned to Pearl after nine days out, only eight other bat-tleships moored to the quays alongside Ford Island. BatDiv 3 had dis-appeared into thin air. We discovered that the Chief of Naval Operations had transferred the *New Mexico, Idaho,* and *Mississippi* to the Atlantic Fleet, along with the carrier *Yorktown,* four light cruisers (the *Brooklyn, Nashville, Philadelphia,* and *Savannah*), two destroyer squadrons, three transports, three oilers, and several other auxiliaries. With one stroke of his pen, Admiral Harold R. Stark had reduced the strength of the Pacific Fleet by at least twenty percent. We were now outnumbered by the Imperial Japanese Navy in every category of combat ships. We did not know that then, and probably wouldn't have worried even if we had. After all, wasn't one American fighting ship equal to two or three of Japan's?

This was only the first in a rapid series of increasingly ominous develop-ments. On 24 May the new German battleship *Bismarck* destroyed the 48,000-ton British battle cruiser *Hood* with a few well-directed salvoes, then bloodied the new battleship *Prince of Wales* and sent her into igno-minious retreat. This remarkable feat gave rise to much discussion, mostly critical of the British. When were the Limeys going to learn how to build capital ships that didn't explode when they got hit? If they couldn't manage this, when were they going to stop trying to use battle cruisers as bat-tleships? At Jutland in 1916, three of that bastard combat type had blown up—now the *Hood.*

There were many chuckles and guffaws as the British sent their entire Home Fleet, plus units of other commands, in pell-mell pursuit of one battleship and her cruiser escort, the *Prinz Eugen.* The traditional Amer-ican sympathy for the underdog was evoked, and many of the crew were rooting for the *Bismarck* to make it safely into Brest, despite the fact that the British were nominal allies and the Germans likely foes. We were unanimously agreed about a couple of things: no American battleship would blow up as the *Hood* had, and the Germans would be worthy adversaries.

We stopped laughing on 27 May. That was the day the British armada finally caught up with the *Bismarck,* after a lucky torpedo hit had jammed the German battleship's twin rudders and sent her turning in helpless circles. It required five battleships, three battle cruisers, two carriers, four heavy and seven light cruisers, twenty-one destroyers, and more than fifty aircraft to send her to the bottom in a Götterdämmerung of horror for nearly 2,100 German sailors.*

*Von Müllenheim-Rechberg, Baron Burkard, *Battleship Bismarck: A Survivor's Story* (Annapolis, Maryland: Naval Institute Press, 1980), p. 243.

Something else happened that day which affected us much more directly. President Franklin D. Roosevelt declared a state of "unlimited national emergency."

For the regulars in the *California*, there would be no discharges when enlistments expired. For the growing number of reserves, including me and the other five V-3s of the radio gang, there would be no release from active duty "for the duration." My new goal of college would have to be deferred for a while.

At first, I was the butt of some good-natured ribbing. "Welcome to the navy, Mason!" "How do you 'feather merchants' like active duty now?" "Hey, written your congressman yet?"

But one regular, himself nearing the end of his enlistment, summed up the new mood best: "Buddy, from now on we're all in the same boat together!"

One very welcome effect was that it increased the solidarity of the C-D Division. The regulars stopped referring to me as a V-3 and accepted me fully into "the gang." And I stopped thinking of myself as a reserve. For better or for worse, it was one navy now.

Despite these portents, it was a navy whose leaders seemed reluctant to abandon their rigidly prescribed peacetime ways. The ships were still painted pale rather than battle gray. The men continued to wear the white shorts and T-shirts (uniform of the day in Hawaiian waters since 26 October 1940) that would result in so many horrific burn cases a few months later. At least a third of our time was devoted to tedious protocols that had little to do with preparation for combat.

Every Friday morning there was a great scrubbing and swabbing and polishing so that the exec would not dirty his white gloves on the top of a locker or a stanchion at the afternoon material inspection. Every Saturday morning, we mustered on the quarterdeck for personnel inspection, a ritual made more onerous by Bunkley's custom of keeping the entire crew standing under the hot Hawaiian sun for an extra half hour or forty-five minutes while he conferred with ComBatFor or ComBatShips or ComBatDiv 2 in his cabin. I used to wonder whether this was conscious sadism on his part, or whether he was simply oblivious to the feelings and the welfare of the crew, and decided it was probably the latter. When he emerged from the quarterdeck hatch with his party, cumulative waves of hostility arose from the dressed ranks, like the miasma of a winter tule fog in Sacramento; but he seemed oblivious to this, too.

With every visit from a VIP, and they were numerous, four, six, or eight side boys were paraded, all personnel on the quarterdeck were called to attention, and the bosun's pipe whistled its shrill notes. Every departure of Admiral Pye, if he was in uniform, called for honors even more elaborate.

All this time the ready ammunition boxes were locked, as if the senior officers had more to fear from their countrymen than from any external enemy, and the location of the keys was usually a mystery.

We soon received yet another sign that, if we would not go to the war, the war was all too willing to come to us. On a typical June day in Oahu, with scarcely a whisper of a tradewind and Pearl Harbor as flat and calm as a millpond, we looked up to see a strange warship make the turn around Ford Island and approach Ten-Ten Dock.

The marine guard of the day and the band were called out. As the bow of the ship came opposite our fantail, the guard presented arms and the band struck up the patriotic song I knew as "America" but which, I also knew as I faced outboard and saluted, was the national anthem of Great Britain. Sure enough, the White Ensign with the cross of Saint George in the upper left corner was flying from her gaff.

She was a six-inch-gunned light cruiser, with a low, boxy superstructure topped by a large director control tower. Her light-gray paint was marred with rust and red lead. She was rather low in the water and seemed to have a slight list from some kind of battle damage. The word soon came down from the signal bridge that she was HMS *Liverpool*.

She looked considerably smaller than her rated 9,400 tons and decidedly unshipshape; but then, we didn't think much of the British Navy. In our minds we saw the *Hood* blowing up in a towering fireball, with a hanging column of black smoke as her funeral pyre. The Italians were the only ones the British seemed to have any success against, and the common attitude was that the cowardly Italian Navy was even more laughable than Nippon's. Against one of our modern 10,000-ton light cruisers, we could easily see, the puny *Liverpool* wouldn't have a chance. It seemed obvious why the Limeys were losing the war. The other "experts" and I hadn't yet learned that combat-hardened crews in mediocre ships can often defeat untested and unblooded crews in superior ships.

Two days later, on 9 June, the *Liverpool* unmoored and slipped out of Pearl Harbor. By now I had heard the scuttlebutt: she had been torpedoed in the Mediterranean, and was en route to Mare Island Navy Yard for repairs. The significance of this blatant violation of our professed neutrality escaped me; I watched her leave with regret, not having had the opportunity to pay her a visit and talk with the crew, particularly the radiomen.

My chance to get acquainted with some British sailors came less than two months later. After a morning watch, I climbed four ladders to the forecastle for a breath of fresh air. Moored to Ten-Ten Dock, almost directly across the channel, was a strange and obviously venerable battleship. She too was flying the White Ensign.

"What ship is that, Mac?" I asked a coxswain.

"Limey battlewagon. The *Warspite*." He looked cross channel, adding with obvious respect, "She was in the Battle of Jutland."

"The *Warspite*! I'll be damned!"

Every schoolboy had read about Jutland, the epic naval engagement between the British Grand Fleet and the German High Seas Fleet on 31 May 1916. Accordingly, I knew about the *Warspite*. Then a nearly new dreadnought, turning in circles with a jammed rudder, she had shaken off salvo after salvo from Scheer's main battle fleet while returning a murderous fire against the German battleships and Hipper's force of battle cruisers. Now here she was at Pearl Harbor twenty-five years later, 30,000 tons of naval history.

She may have been past the age of retirement, but she still looked formidable. She had more freeboard than American battleships—a feature that was emphasized because she was riding light and showing red below her black boot-topping—and fifteen-inch guns in four paired turrets. In place of her original, graceful tripod foremast, she had been fitted with a high, solid tower superstructure and midships aircraft hangar and cranes (during a mid-1930s rebuilding, I soon learned). Her gray paint was stained and blistered so that, though not showing any gross signs of damage, she looked weather-beaten and battle-scarred. The *Warspite* somehow reminded me of an old ex-heavyweight champion, far past his prime, who could still take a punch and still had a mighty right hand.

The *Warspite*'s past was indeed impressive. At Jutland she had been with the fast battleships *Barham*, *Valiant*, and *Malaya*, when the grand fleet commander, Admiral Sir John Jellicoe, was, as Churchill said, "the one man who could have lost the war in an afternoon." What tension, what drama, what deeds of valor her decks had witnessed! The officers and men who had fought aboard her that fearful day and night off the Danish coast were now old men—or dead. Another generation was treading in their footsteps, operating the same machinery, turning the same rudder from the same wheel, firing the same fifteen-inch main battery that had roared its defiance of the Hun so long before.

To my delight, Lieutenant Sugg asked me if I wanted to be one of the enlisted hosts to a visiting party from the *Warspite*. The young sailor I shepherded around the *California* was a leading seaman, the rating equivalent to a third-class petty officer in our navy. He was slim, delicately handsome, and self-possessed, projecting the faint aura of British superiority that many have found so irritating about Churchill's "stout island race."

I probably indulged him a bit, since ancestors on both sides of my family had come from Britain. Besides, he was from the fabled *Warspite*, a fact that alone made him special. I was anxious that he receive a good impression of my navy and my ship. I wanted him to see for himself that Americans were not all cowboys and gangsters and hooligans.

My guided tour included the tiny library, just off the crew's reception room, whose shelves were stacked with books written by Americans. The chaplain's yeoman found a copy of Bunkley's *Military and Naval Recognition Book*. "Written by our skipper," I said pridefully.

He arched an eyebrow. "Really?" He leafed through it politely and handed it back. "Not much use to you and me, is it?"

Our next stop was the gedunk stand, where I planned to buy him a Coca-Cola. That should be a real treat after the sarsaparilla that, I had heard, the English were forced to drink. To my surprise, he rejected the Coke out of hand.

"You chaps don't have any grog on your ships, do you?" he asked.

I explained that U.S. Navy ships had been dry since a sanctimonious Puritan named Josephus Daniels had been secretary of the navy during World War I. His lip curled slightly. I could see that his opinion of Americans as barbarians had changed but little despite my propagandizing.

"No wonder you Yanks get a little boisterous ashore," he commented sympathetically.

My offer of a gedunk was also rejected with barely disguised horror.

"Tea?" he asked. I could well imagine the reaction of the seamen soda jerks, surly fellows at best, should I dare ask for a pot of tea. I explained that coffee was available in unlimited quantities, but tea was rarely served.

Clearly my guest felt that he might as well be visiting a ship of the Turkish Navy for all the amenities offered in the *California*. But with the good manners one associates with the British, he accepted a cup of navy joe, heavily sweetened with sugar and canned milk. His one other concession to our culture, and that a happier one, was the purchase of a carton of American cigarettes, for which he insisted on paying the sixty cents.

Our tour took us through the gun casemates and ended on the boat deck, where I showed him our five-inch twenty-five-caliber and three-inch fifty-caliber antiaircraft batteries.

"What do you think of my ship?" I asked hopefully.

"Not enough ack-ack," the Briton answered tersely.

I was truly surprised. I thought our boat deck was bristling with deadly weaponry designed to bring down attacking aircraft.

"What about all these?" I asked, indicating the slim, long-barreled three-inch and stubby five-inch rifles in their unshielded mountings.

"Not enough by half," he replied. "We have a good deal more than you do, and it still wasn't enough at Crete." He explained that the *Warspite* had been hit by a 500-pound bomb from a Stuka dive-bomber when she was opposing the German airborne invasion of Crete in May. Thirty-eight men were killed and another thirty-one were wounded.

I knew he was speaking sincerely, not merely from British arrogance.

This being a good-will visit, I didn't wish to argue, though I disagreed with his opinion of our antiaircraft defenses. How could a mere leading seaman in the Royal Navy know more than all my gold-bedecked superiors? After all, Americans had a special genius with weapons, just as they did with machinery. Weren't English lads playing cricket or soccer while American boys were bringing down game with their .22s? I couldn't imagine the young men of the two countries being anywhere near comparable in ability at handling firearms of whatever type. I may have been half right, but my guest in the *California* that day proved considerably less than half wrong.

Before we parted, he invited me to visit him in the *Warspite* the next day. I accepted with alacrity.

As I stepped from the after brow onto the quarterdeck of the *Warspite*, it seemed almost un-American to be saluting a foreign ensign and a foreign uniform, but I told myself I was actually saluting the ship's glorious record. The officers did not seem as nattily uniformed or as taut as ours, but they more than made up for these deficiencies with their innate civility. Strange, I thought, that our officers, who had borrowed so much from the British, seemed to surpass them in hubris.

As he showed me about the ship, my guest of the previous day spent a great deal of time in the crew's messing and berthing compartments, as if there were a special message there for me and for my navy. Unlike the centralized galley facilities in our ships, each division had a separate mess and its own cooks.

Although it was verging on mid-afternoon, the tables were still down and the men were taking their ease, eating and drinking tea. Some of them were enjoying something stronger than tea, which would have been a court-martial offense in the *California*. By American standards, the ship was dirty and the bulkheads badly needed painting; the *Warspite* was showing her years. There was a general air of informality verging on disorganization. The men were older than the crew of the *California*, and simultaneously insubordinate and respectful in what I later learned was the typical British lower-class style. I was introduced to some of them and they made amiable if slightly sardonic conversation, looking at me across the light-years that separate men who have been in combat from those who have not.

Later, we passed through officers' country, something we could not have done in my ship. Rich walnut paneling had been polished over three decades to a deep, mellow luster, and staterooms were filled with comfortable chairs and books and the kind of casual disorder that reminded me vaguely of what I had read about Oxford and Cambridge. I felt sure it would not be out of order to quote Shakespeare or Chaucer, Wordsworth or Housman in these staterooms, as it would have been in the *California's*.

That kind of knowledge was undoubtedly expected of an officer and a gentleman, according to my interpretation of the British view. I promptly became an anglophile.

As we passed one stateroom, an unusually young, skinny attendant was emerging, carrying a water jug or a thermos bottle. I stared, as perplexed as my host had been the previous day in the gedunk line. The boy had scarcely entered puberty.

"Do you mind telling me how old he is?" I asked.

"Thirteen, perhaps fourteen," was the casual reply.

"And they let him in the navy?"

"Of course. He's a cabin boy."

For such a recent convert to the British cause, it was rather a shock to discover a thirteen-year-old cabin boy serving in HMS *Warspite* in the middle of a war. Especially when I reflected that this pale, gawky adolescent, still too young to shave, already had more combat experience than anyone in the U.S. Navy.

While my host and I were parting, I stared up at the high, angular superstructure of the rusty and wounded *Warspite*. She was still 30,000 tons of naval history. I wished him luck.

"I won't be needing that for awhile," he said, again with his faint and invincibly British smile. "We're going someplace near Vancouver—Puget Sound, is it?—for repairs."*

I told him about the navy yard and about Bremerton and the places he should visit in Seattle.

"Yes," he said. "I'm going to be there, in your country. And you're going to be out here in this bloody awful hole, with not half enough ack-ack. You're the one who needs the luck."

He was right about that, too.

---

*The *Warspite* left the following day (4 August). She arrived in Bremerton 11 August and was still there on 7 December 1941.

# CHAPTER 9

# Summer and Gunsmoke (And a Few Training Casualties)

"What we got here is a seagoing goddam boot camp!"

I was leaning against the lifeline at the break of the boat deck, watching a party of about twenty-five apprentice seamen report aboard from the San Diego Naval Training Station, when a familiar bellowing voice announced the arrival of Chief Radioman Reinhardt.

"If they can't train these gedunk sailors at Great Lakes or Dago, why don't they ship 'em to the Hooligan Navy—the goddam flattops?" Reinhardt demanded. "It don't make no difference there!"

I agreed with the flag chief. All summer, parties of raw recruits had been checking in, looking apprehensive and ill-at-ease as they lined up on the quarterdeck with bags and hammocks. And just as regularly, experienced seamen and petty officers were being transferred off for school instruction, for new construction, and for duty in ships of the Atlantic Fleet.

"The next thing you know, they'll be shipping us a bunch of Red Cross broads, so these puling kids won't miss their mamas!" the chief shouted in red-faced disgust.

To most of the enlisted men, it looked as if Kimmel was ignoring the battle readiness of the Pacific Fleet in the interests of a vast on-the-job training program. A few old hands pointed out that the Bureau of Navigation was probably directing the transfer of trained personnel to the North Atlantic Sea Frontier, where we were in an undeclared shooting war with the Nazi U-boats. Kimmel, they said, was undoubtedly "bitching and moaning" to CNO about losing so many of his key petty officers.

The Pacific Fleet had been reorganized into three task forces. They were still, essentially, the Battle Force (Task Force One, under the command of Vice Admiral W. S. Pye in the *California*); the Aircraft Battle Force (Task Force Two, under the command of Vice Admiral W. F. Halsey in the *Enterprise*); and the Scouting Force (Task Force Three, under the command of Vice Admiral Wilson Brown in the *Indianapolis*). But Pye had lost three of his nine battleships (the *Colorado*, *Maryland*, and *West*

*Virginia*) to Halsey, and Halsey had lost two of his three carriers, one to Pye and one to Brown, so that each task force could be screened by both destroyers and carriers against enemy submarines. For some reason, Kimmel seemed obsessed with Japanese subs.

Generally, each task force went to sea on training maneuvers for one week and then was moored in Pearl Harbor for two weeks. Occasionally, a simulated fleet engagement brought two or all three task forces to sea; but this was as rare as the sight of Kimmel's four-star flag flying from the *Pennsylvania* at sea. Usually, he ran the fleet from the administration building at the sub base.

That summer the tempo of international events quickened. On 22 June Hitler unleashed the staggering total of 120 divisions against the Soviet Union along a 2,000-mile front stretching from the Baltic to the Black Sea. Roosevelt promptly promised American aid to the beleaguered Russians, a move not greeted with unanimous approval in the fleet. Many career enlisted men, conservative by nature, saw the invasion as a heaven-sent opportunity to destroy the hated Bolshevik dictatorship once and for all. If they had had their choice, they would have sent aid not to the Russians, but to the Germans.

On 2 July Japanese military forces took over French Indochina from the feeble Vichy regime. On the seventh U.S. forces landed in Iceland to relieve British troops stationed there. And on the twenty-sixth Roosevelt halted trade with Japan, froze Japanese assets in the United States, and nationalized Filipino military forces under General Douglas MacArthur, remembered unfavorably by many as the army commander who had turned his cavalry and infantry troops against the "Bonus Army" of jobless World War I veterans in the summer of 1932.

In August the big event was the four-day meeting of Roosevelt and British Prime Minister Winston Churchill on board the heavy cruiser *Augusta* and the battleship *Prince of Wales* in Argentia Bay, Newfoundland. The result was the Atlantic Charter, which called for the "self-determination" of all nations and the defeat of the Axis powers. Once again, according to the political analysts in the petty officers' head, our "idiot leaders" in Washington were being maneuvered into "saving the ass of the Limeys."

Less than a month later, on 11 September, Roosevelt issued his famous "shoot-on-sight" order to any U.S. naval forces that discovered Axis vessels in American waters. In effect, the order applied only to our Atlantic Fleet and was directed against the German Navy. Again, the petty officers groused about "Franklin and Eleanor being in the sack with Winnie." Damn it, the real enemy was out here in the Pacific!

All these momentous events seemed utterly remote from Task Force

One as it repeatedly cruised the seas south and west of the Hawaiian Islands.

"You see those grooves out there in the water?" one bosun's mate asked an impressionistic boot. "We've crossed this area so often we've worn tracks in the frigging ocean!" While the recruit scanned the waters abeam, the bosun's mate winked at me. "If we stay out here much longer," he said sourly, "I'll be seein' 'em, too!"

My training program was beginning to pay dividends. I was now entrusted with the NPM Fox sked, with a striker or newly arrived V-3 as my back up. Occasionally, I was assigned to a circuit that called for a "live" transmitter, even though I seldom had a chance to use it. I was also sitting in on the training circuit controlled by the ComBatShips flag radiomen in the *West Virginia*, where I had plenty of opportunities to key a transmitter.

This circuit was not without its hazards. It was monitored by senior firsts, who copied down everything that was sent, right down to the number of *dits* (letter *e*'s) in a "bust." Soon copies of their logs would be posted on the bulletin board in main radio for Chief Reeves to analyze and for the entire radio gang to chortle over. My embarrassment was mitigated only by the discovery that most of the other thirds were making just as many mistakes as I.

One day Reeves called me over to the supervisor's desk. He was holding a new watch, quarter, and station bill. "Mason, I'm giving you a promotion," he said. "I'm sending you to the maintop for your battle station."

"That's great, chief!" I enthused. "What do I do up there?"

"When our planes are up, you copy their spotting reports," he said. "By hand. They're used for correcting main-battery fire."

"What do I do when the planes aren't up?"

"Enjoy the scenery." He smiled. "You're going to like it up there, Mason. Lots of fresh air."

The chief had given me more than a promotion. He had, it is quite likely, given me my life. Otherwise, I would have been in main radio on 7 December. After it was abandoned, I would have been with Johnson near ship's service when a bomb hit there. Or—had I been a better man than I probably was—with Reeves in a burning passageway on the third deck.

At the next call to general quarters, I raced toward the quarterdeck and up a series of steep ladders inside the cage mast to the exposed "bird bath" at the very top. There, on a narrow catwalk encircling the main-battery director, protected only by a thin, chest-high splinter shield, I plugged a headset into a jack and stood by with message pad and pencil to copy the reports on the fall of shot, transmitted in weak and wavering signals from our OS2U Kingfisher observation planes.

The author's battle station was in the exposed "bird bath" encircling the main-battery director at the very top of the mainmast, where he copied radio reports on the fall of shot from the ship's observation floatplanes. About the only discernible differences between the *California* in this April 1935 photo and in the summer of 1941 were the addition of "bedspring" antenna for the CXAM radar atop the superstructure and the replacement of biwinged O3U Corsairs with mono-winged OS2U Kingfishers. (Photo courtesy of the National Archives.)

If I copied "KU 200," for example, it meant that the 140-foot-long by 40-foot-high latticework target raft had been straddled with in-line splashes from our fourteen-inch guns, mostly falling a little short, and that the elevation should be raised sufficiently to increase the range by 200 yards to score a bull's-eye.

Reeves had been right about the scenery and fresh air. From the maintop I had a panoramic view to the horizon, obstructed only toward the ship's head by the foremast one hundred feet away. In the foretop I could see Fisher, who had also been promoted and was copying the same spotting reports that I was. The chief had neglected to tell me, however, that the mainmast was just abaft the number two stack. When we steamed into the wind, the stack gas would blow back and envelop us in its reeking hot effluvium, and my white uniform would soon be covered with greasy black polka dots of carbonized fuel oil.

Better that, I thought, than the choking confines of main radio below the dogless hatch cover. Now I was as close to the big guns that were the *California's raison d'être* as a radioman was ever likely to get. And, for the first time in my navy service, I was operating without direct supervision—a vote of confidence I had worked hard to earn.

On the bird bath with me were the lookouts with their binoculars, a talker on the sound-power phone, and the crews of the four fifty-caliber machine guns. Inside the three-level director (main-battery director and control station in the top two; secondary-battery director and control below) were the fire-control personnel. To prevent blast damage, the tiers of glass windows in the director were opened during general quarters.

My nominal superiors were Ensign Benjamin C. Hall, F Division officer in charge of Spot II, to whom I passed my OS2U reports, and Ensign Edward R. Blair, Jr., fresh from the U.S. Naval Academy, in charge of the Sixth Division machine gunners. Hall, Annapolis '40, I remember as a brisk, no-nonsense officer who coached the ship's basketball team. Blair I clearly recall, principally because he had a broad "magnolia" accent, delivered in a husky voice, and treated me as if I were a cotton chopper on the plantation.

Firing the *California's* twelve fourteen-inch rifles was an incredibly complex procedure, one that directly involved nearly 500 men and could only be mastered by constant drill. From magazines and handling rooms on the first platform deck, shells and powder moved by electric hoist up through the barbettes to the gun rooms. There the projectiles were loaded hydraulically into the breeches of the guns. Behind them went two-toned silk powder bags that contained black ignition and smokeless propelling charges. Last came a cartridge primer designed to fire into the volatile black powder placed in the red end of the bags. When the massive breech mechanism with its quick-closing interrupted-screw plug was locked, the guns were ready to be fired electrically from any of several locations.

Before this could happen, the guns had to be aimed. Large optical range-finders, located in the foretop and maintop, atop the superstructure, and in the turrets, where they projected out from the sides, established the distance to the target. The stabilized telescopes in the main-battery directors in the battle tops determined the target's bearings from the ship. The range and bearing information was fed to the plotting room, located just abaft central station on the first platform, which housed the scientific marvel of that day: a Ford analog fire-control computer, the first ever manufactured.

The preliminary data from the range-finders and directors only revealed where the target was at that moment. To score a hit, it was necessary to predict, by rapid solution of multi-termed mathematical equations, the location of the moving raft or vessel at the end of the projectiles' flight, again compensating for any ship's motion. The computer calculated the train and elevation angles necessary to intercept the target, taking into consideration such factors as the effects of wind, temperature, and barometric pressure on the projectiles, and even the effect of the earth's rotation upon their trajectories. The resulting fire-control calculations were transmitted electrically to the turrets, where the pointers and trainers read the indicators and made the necessary corrections.

When the twelve "gun-ready" lamps in the turrets, conning tower, plotting room, and fire-control towers glowed red, the order to shoot sent the 1,500-pound projectiles on their way. Only if the men and their equipment had done a near-perfect job would the result be a straddle around the target.

After the first, or spotting, salvoes, while powder and shell continued to move up from the magazines for the next salvoes, spotters in the tops estimated the fall of shot and relayed corrections. If the spotting planes had been catapulted, as they usually were except for night and short-range battle practice, Fisher and I copied and passed along their corrections. The computer digested this information and cranked out the new train angle and elevation for the gunlayers in the turrets. To aid the spotters in the tops and the observers in the seaplanes, the shells carried brilliantly colored dyes that marked the splashes in macabre fountains of blue, green, orange, and vermillion—a color for each battleship.

Seen and heard from the open battle top, the firing of the fourteen-inch guns was terrifying. Mighty thunderclaps of sound. Yellow-orange flames eagerly licking up toward us, followed by clouds of light-brown smoke which smelled as if it had just come from the nether regions. Vibrations that shook and rattled the mainmast. The convulsive lurch of the 35,000-ton ship as she was slammed sideways in the water. The rumble and roar of what sounded like high-speed locomotives when the three-quarter-ton projectiles rose in high sweeping arcs toward the horizon, glowing like bolts from hell.

The spotters and gun crews were allowed to stuff cotton into their ears for a modicum of protection against a din that sounded like the onslaught of Judgment Day. I, of course, could not. And the earphones were worse than useless, only seeming to amplify the sound to near-intolerable levels. I wonder to this hour why I didn't suffer permanent hearing impairment from those days and nights at the parapet in the maintop.

But in the summer of 1941 I had something else to wonder about: what if those bolts from hell were coming my way at more than 2,500 miles an hour? The *California* was protected against plunging shellfire on the all-or-nothing principle. If the projectiles hit where it counted—the magazines, turrets, engine-room spaces, conning tower—they would be turned aside by up to eighteen inches of armor plating. If they hit where it didn't count—areas unprotected by armor, like the maintop—then it didn't matter anyway. Except to the people who were stationed there, I thought darkly. Then I was informed that only a highly specialized fuse would explode a shell before it had penetrated and passed on through the thinly protected battle top. The principal danger, I was told, was bursting flame and shrapnel from hits below, on the main or boat decks.

That was but cold comfort. I knew my reflexes were excellent—but hardly good enough to duck a fourteen-inch projectile should it choose to occupy the space where I was standing.

With the Pacific Fleet consigned to Pearl Harbor for the indefinite future, the navy made some token efforts to provide for the recreational needs of tens of thousands of enlisted men, who had few outlets for their energies beyond the bars and brothels of Honolulu. The fleet recreation center at Nanakuli on the western coast of Oahu was one such effort. Using a little of the swollen ship's service fund, the *California* periodically sent large liberty parties on weekend excursions to Nanakuli.

When C-D Division's turn came in July, Johnson and I signed up, more from lack of funds than in expectation of a memorable time. That it turned out to be memorable indeed was hardly the result of advance planning.

The Nanakuli party assembled around 1100 on the port quarterdeck with swimming suits and the minimum toilet articles required for an overnight stay. By special dispensation we were allowed to wear dungarees. Since the party included one of the deck divisions, as well as the radiomen and signalmen, it was a large one: nearly one hundred men. An ensign was in charge, assisted by two or three boatswain's mates representing the master-at-arms.

At Aiea we boarded the Oahu Railway's Toonerville Trolley, named after the rattletrap rural line featured in a popular comic strip of the 1930s. The nickname was exquisitely appropriate, since the ancient narrow-gage cars were equipped with slatted wooden seats as unaccommodating of the human derriere as park benches, and offered natural air-conditioning through permanently open windows. Huffing and puffing and emitting showers of sparks, the Toonerville passed through the sugarcane planta- tions of Ewa and skirted the southern edge of the Wainae Range before making its way up the leeward coast of the island.

Nanakuli proved to be a wide, white-sand beach fronting a gentle ocean whose surf was little higher than Lake Tahoe's and a great deal warmer. It was a fine safe beach for inexperienced swimmers, which most of my shipmates were. Dusty Camp Andrews was located across a two-lane country road, among the coconut palms. The Royal Hawaiian Hotel it was not. Nondescript wooden buildings housed an army-type field kitchen and barracks with GI cots. The heads, likewise, were army latrines.

There was little to do but swim, body surf, play touch football in the sand, and drink beer. Galvanized laundry tubs partly filled with ice served as effective coolers for limitless quantities of beer. Discovering that it was not Oahu's vile Primo beer but a popular California brew called Acme (the first beer I had ever drunk, a year before while camping out in the El Dorado National Forest with Fisher and his family), my buddies and I turned to with a will.

Late afternoon chow was hot dogs, cold cuts, beans, and cole slaw, accompanied by more Acme beer. By then, a huge yellow-orange sun was extinguishing its fires in the Pacific Ocean, and the advent of the Hawaiian night was reminding even the young of their mortality. Breathing that pure and intoxicating air, faintly perfumed with blown spray, yellow ginger, and frangipani, I began to understand the unrestrained sensuality of the origin- al inhabitants. That pagan Hawaiian night seemed to bring with it freedom from the sober responsibilities of sunlight and for indulgence in all of man's pleasures and vices.

Now someone lighted a huge bonfire and all of us, deck hands and communicators, gathered around it. Not in one jolly group, of course, but by clans. The deck divisions considered the radio gang effete (they would have used a different word) and sneeringly referred to us as radio girls. Confident of our superior skills and education, we in turn thought the members of the six line divisions gross and stupid, and called them deck apes. Across the fire the clans eyed each other coldly.

With even a minimum of supervision, nothing would have happened. Bowing to navy discipline was by now almost instinctive with most of us, ar d the mere presence of an authority figure evoked memories of the stern re tribution that swiftly followed misconduct. This night, however, there was no such presence. The ensign had disappeared hours before. Nor were tl e envoys of Master-at-Arms Spud Murphy around. I was sure they had not been able to resist the beer and had so freely indulged that they were sound asleep somewhere.

Lacking supervision, all such a gathering needs to become a free-for-all is a catalyst. And we had one. His name was James K. Evans, a signalman striker in C-L Division. He was all of eighteen or nineteen, a tall thin Texan with a fair complexion, lank brown hair, and a long slightly hooked nose. I had seen him a few times before, on the bridge or in the signalmen's compartment just forward of my own.

Suddenly he was on his feet, waving a bottle of beer that was by no means his first. "I'm gonna do an Indian war dance," he announced in a thick, slurred voice. Cheers from several of his companions greeted this declaration. Tossing his bottle into the fire, where it sizzled and shattered, he began a rhythmic solo movement.

Johnson and Goff and Stammerjohn and I laughed and clapped our hands in approval. Evans was doing a real Indian war dance, if I could judge by Hollywood westerns. His bare feet slapped the dirt in a seemingly simple but intricate pattern. He was wearing only a pair of dungarees, and his lithe tanned body bent and pulled back and bent again in time with his shuffling feet. His arms, crooked at the elbows, fanned the air in a precise accompaniment. On his face was a remote, intent look, and it did not take much imagination to project oneself back a hundred or a thousand years to another campfire and another group of braves preparing to take the war-path.

"Look at that goddam Papoose go!" shouted one signalman. Chief was the usual nickname for American Indians in the armed forces, but Evans was too young and looked too Anglo-Saxon to be a chief. So he was called Papoose, a nickname that stayed with him for the rest of his navy service.*

He danced for a long time, until his body was glistening with perspiration, and we whistled and stamped and cried for more. At last, a deck hand was moved to a different kind of action. Leaping up from the fire, he grabbed Evans by the arm. I didn't catch all the conversation, but his meaning was clear: he wanted Papoose to knock it off.

This elicited mutterings from C-D and C-L divisions. Papoose broke loose and staggered away to resume his war dance, pursued by the seaman. Quickly a radioman—I believe it was the truculent Red Goff—was on his feet.

"Leave him alone, you frigging deck ape!" he bawled.

Goff then turned to his messmates with a call for battle stations: "Do we let him get by with that shit? I say negative! Let's go!" He charged around the campfire toward the seaman, followed by four or five of the larger and more aggressive (and perhaps more intoxicated) radiomen and signalmen. By now, most of the deck hands were up and met Goff's forces head on.

*From late 1942 to early 1945, I served with the late Jim Evans in the *Pawnee* (ATF 74), principally in the South and Central Pacific. He was a fine shipmate and a man of valor.

But I had no time for further observation. A group of seamen was making its slow way toward us with mayhem obviously in mind. Even then, Johnson—the man who prided himself on being "a lover, not a fighter"—took one last swig of his beer before flipping it away. "Stay close to me, men," he said, grinning crookedly.

My hastily formulated plan was to do exactly that. Not long before, in a North Beretania Street bar, a former navy boxing champion known to all as Tiger had offered a half dozen of us younger sailors some practical advice. "In a barroom brawl," he said in an adenoidal fighter's voice that still, after three or four "hitches," bore traces of his Louisiana antecedents, "the first thing you gotta do is protect yourself. If you're alone, try to fight your way to the door. If that ain't possible, find a corner where they can't get at you from behind and let 'em come to you.

"Now if you got a buddy with you it's a lot better, 'cause you can protect your flank. Find your corner and get up against each other, back to back, facin' out a little." Using me, he showed how this V-formation made it very difficult for more than one opponent to get at us. "Now with this system," he concluded, "me and Sparks here could take any six guys in this bar." As we said goodnight to him later at Merry Point, I very much regretted that the broken-nosed Cajun was not serving in the *California*.

Especially at this unexpected and unwelcome moment. Now I was forced to put Tiger's self-defense tactics to the test with neither the benefit of his presence nor a friendly corner. I realized, with a sick feeling in the pit of my stomach, how stupid it was to fight my own shipmates, when we had so many real, external enemies. But it would be even more stupid not to defend myself.

Back to back with Johnson, chin tucked behind my left shoulder, I tensed for the assault. Tiger had correctly predicted that the first seaman to reach me would throw an overhand right. The fist coming toward me looked hard and menacing. I instantly realized that I had forgotten to ask Tiger about a minor technical point. If I moved toward my opponent and let the punch go by, there was no wall to absorb the impact, only Johnson's head. Obviously, I had to deflect his arm before countering, a maneuver that required precise timing.

Instead of delivering a neat sequence of body hooks, as Tiger had counseled, I found myself in a wild exchange and took a few glancing blows: proof that nothing ever quite goes by the book. Still, by staying inside and focusing my counterattack on the midsection, I forced the seaman to give ground. He turned away, grunting in discomfort.

I was better prepared for the next one, parrying with the left and countering with the right in the same motion. Thanks as much to the Acme Brewing Company as to my skill, the fellow sagged to the ground, retching quantities of beer. At about the same time, Johnson landed a thundering right to the head of his tall attacker, sprawling him full length on his back.

The melee swirled past us like waves parting around a rock as the seamen sought easier prey. Tiger had just made a true believer of me.

Johnson grinned and shook out a skinned right hand. "These deck apes sure have hard heads, buddy! Let's see if anyone needs help."

The brawl had disintegrated into a series of isolated skirmishes. Flushed with victory, we headed toward a cluster of hard-pressed radiomen. But at last the peacemakers had been roused. The bosun's mates ran back and forth, waving their night sticks and spewing obscenities. The ensign was doing a gingerly dance of his own among the combatants, brandishing his .45 and admonishing us in a high, plaintive voice to break it up.

"That's an order! Do you hear me, that's an order!"

Gradually, the two forces were separated and stood in loose, panting groups, glaring at each other across the campfire.

The last to rejoin C-D Division was Red Goff, looking flushed, slightly scarred by combat, and triumphant. He had had the first word, and now he couldn't resist the last one.

"Hey, you deck apes!" he shouted. "Whatta ya think of the radio girls now?"

The deck apes had, clearly and surprisingly, lost the battle. Johnson met with much approval for his thunderbolt right—"the hammer of Thor," he described it with becoming modesty. Even scrawny little Stammerjohn was a hero, having landed a Sunday punch on an unsuspecting but much bigger foe. It was probably the only blow he had ever thrown in anger in his life.

With some difficulty the forces of law and order, now wide awake, kept the truce through the night and en route to the ship the next morning.

"We owe you one," a bruised and scowling seaman muttered as he passed down the aisle of the Toonerville Trolley. "You radio girls better stand by."

A middleweight radio striker, who happened to be an amateur boxer and had distinguished himself during the brawl, came out of his seat. "How about right now?" he challenged. An encore was narrowly averted.

So in that tense, hot summer of 1941, with half the world at war and our admirals emphasizing the Pacific Fleet's training mission at the expense of its combat readiness, the radio gang of the *California* won a grudging respect. Even Chief Reeves chuckled with satisfaction over the results of the donnybrook at Nanakuli; and I entertained illusory hopes that Johnson would at last get off the "shit-list."

Shortly afterward, one of the seaman who had been involved in the brawl was observed loitering near the main radio hatch, belaying pin in hand. The master-at-arms was called and he was disarmed. Some in the radio gang found our humiliation of the deck hands amusing. But the threat of a belaying pin to the head as I emerged from main radio at 0100 some morning, with only the dim blue battle lights illuminating crew's space

A-518-L, was not my idea of a joke. It was all too reminiscent of the flying bottles at The Breakers in Seattle.

There were no repercussions of our brawl with the deckhands. The incident was ignored. Most officers, I was told, expected sailors to fight at ship parties and looked indulgently on the practice. They felt it did the crew good to "let off steam."

I could think of much better ways of letting off steam than getting drunk and fighting with my shipmates. Unfortunately, the facilities for healthy discharge of the accumulated tensions of living in close quarters with too many men, such as inter-division and inter-ship athletics, were grossly inadequate at Pearl Harbor.

As the days passed in slow procession, the stultifying routine began to exact other tolls. Most of the casualties were minor: men fell down ladders under darkened-ship conditions, or bruised their heads against unyielding hatch covers while double-timing to their battle stations, or smashed their fingers while operating loading machines.

Some of the incidents were amusing, at least in retrospect. One night the ship was moored to interrupted quay F-3. Precisely at midnight a seaman second, according to the ship's log (27 August 1941), "while walking in his sleep, jumped overboard from starboard galley deck; sent to Sick Bay, examined by Lieut.(jg) W. R. Miller, (MC) U.S. Navy, and admitted to sick list. No apparent injury noted." While the log gives no further details, it is likely that the recruit was examined most carefully indeed. Feigning sleepwalking could be an ingenious ploy for getting shore duty or even a medical discharge.

In an incident involving another second-class seaman, no ploy could be detected. This clumsy fellow, "while spreading the quarterdeck awning, fell from the right gun of Turret #4 and sustained contusion of right buttocks and possible incomplete fracture of right humerus. Right arm X-Rayed and put to bed for rest and observation [sic]" (ship's log, 17 August 1941). His rest, assuming the seaman was attached to his arm, was a short one. Three days later he was on his way to the *Oklahoma* for duty. When "eight balls" were culled from the crew of the proud Battle Force flagship, they were often sent to the "Okie," which had no admiral on board to protest. (Another favorite ship of exile was the *Arizona*.)

We used to wonder when our Nazi radio striker would be among the deportees. He was a tall thin seaman second with dirty blond hair and a German surname. On his left wrist was a large circular indentation from the surgery that had removed a swastika tattoo. The operation was performed so he could enlist in the navy, he had told me one day at Bremerton. His Nazi sympathies were well known; he spouted *Mein Kampf* propaganda to anyone who would listen. But no one in the radio gang listened. Since he was periodically in trouble for such offenses as AOL, failure to obey orders, and lying, and had a disagreeable personality, he was considered the rankest of foul balls. Why Chief Reeves continued to tolerate him was something of a mystery.

It was the misdeeds of the admiral's marine orderly that drew the loudest and most sardonic response from the crew. He was put on report for quitting his post at the door to the admiral's cabin without being properly relieved and for being asleep on duty in the early morning hours of 24 September.

Vice Admiral Pye, it should be explained, hardly looked the part of commander of a mighty armada of six battleships, one carrier, five light cruisers, eighteen destroyers, and five mine vessels. He was very short and rotund, with a round, cherubic face below thinning gray hair. Physically, he would have been ideal typecasting for the role of the well-intentioned but bumbling friar in *Romeo and Juliet*. In fact, he was known among the more irreverent enlisted men as Sweetie Pye. So when the sleeping sentinel's deck court-martial was published, there were loud guffaws in the berthing compartments.

"God damn. Someone might have molested Sweetie Pye!"

The only one who didn't appreciate the joke was the admiral's orderly. He got twenty days in the brig.

There was no laughter late one morning in June. As we were preparing to leave our quarters for dinner, we heard men shouting "Gangway!" Four pharmacist's mates came hurrying by, bearing a Stokes stretcher, followed by the ship's junior medical officer. I recognized the fey, pale face on the wire stretcher. It was Wilson, a young seaman in the Fourth Division, a tall, shambling country-looking boy of eighteen or nineteen whom I had often seen in the passageway outside compartment B-511-L. Following emergency treatment in sick bay, he was transferred to the naval hospital, where he died that evening of a fractured pelvis bone and internal injuries. The unfortunate Norris Rayelle Wilson had been caught between the turret mechanism and a stack of projectiles while the turret was being trained.

His death briefly cast a pall over the radio gang. "He's only the first," our philosopher second class predicted in sepulchral tones. "There'll be more."

The next month another tragedy, this one within C-D Division, was narrowly averted. One of our young strikers had lost his twin brother when the latter received a medical discharge at Long Beach many months before. The striker had become increasingly morose and withdrawn, but few were aware of the depths of his despondency, for he apparently confided in no one. On 2 June he received a summary court-martial for being AOL for a full day; and on the nineteenth he was confined to the brig for ten days, with a twenty-dollar loss of pay. That, it seems, was the culmination of what he felt to be a long series of injustices and humiliations.

On 28 July, the day after my birthday, I decided to visit friends in the forward radiomen's berthing space. It was actually the anchor windlass room, a large compartment that occupied the full width of the ship just forward of number one barbette on the third deck. It had been converted to a living compartment—a very noisy one when anchors were being weighed or let go—by sandwiching tiers of bunks and lockers around the three electric-drive windlasses, one for each anchor. A long table and a few chairs completed the furnishings. As I stepped into the compartment, I was greeted by a tumult of excited voices.

"Our crazy striker—— tried to hang himself!"

"A. Q. and a couple others cut him down!"

"Spud Murphy just hauled his ass off to sick bay!"

During noon chow, I was told, the striker had found a secluded corner of the compartment behind the starboard windlass. Securing a line to the overhead, he had put his head into a roughly fashioned noose and stepped off a chair. He was hanging there limply, slowly turning purple, when A. Q. Beal, our dashing second-class radioman, poet, and man-about-Honolulu, arrived on the scene. Shouting for reinforcements, the quick-witted, powerfully built A. Q. held the striker up, taking the tension off the line. Someone produced a knife and cut him down. He was shortly on his way to the hospital and, we heard later, a "section eight" medical discharge.

A bizarre note capped the incident. Around his neck the striker had hung a crudely lettered sign that summed up all of his despair: "I'm just an enlisted dog."

The melodramatic events of that day almost obscured the fact that the law of averages finally caught up with my friend Red Goff. He was given three days' solitary on bread and water for "insolence." That was not surprising, since the truculent hero of the donnybrook at Nanakuli was a master of bold and studied insolence. What was surprising is that Goff had not got himself into truly serious difficulty by decking some arrogant junior officer. Or senior officer, for that matter. Red Goff was most democratic in his attitude toward officers. He despised them all.

Still, most of the captain's masts and courts-martial resulted from AOLs and violent incidents in Honolulu. Typically, the more worldly sailors started their liberties by renting a bicycle or a bathing suit and surfboard in the basement of the Waikiki Inn. After a couple of hours swimming on the beach or pedaling around Kapiolani Park, it was time for drinks on the terrace of the Waikiki Tavern. (Why, I used to ask, had a steep-roofed, timbered and stuccoed Alpine lodge been built on the sands of Waikiki? But no one knew or cared.) As the sailors worked their way back toward

Many Honolulu liberties started with bicycling at Kapiolani Park. From the rear are Bill Fisher, Johnny Johnson, Dave Rosenberg, "Red" Goff, and the author.

Hotel Street, the itinerary usually called for a drink at every bar. The last stop of the evening was the New Senator or one of the other GI houses.

Such liberties were the true test of health and stamina; only the fit survived. Every block held its dangers, in the form of shore patrol, police, or bouncers; merchants, bar girls, or cabbies; occasional roving bands of island toughs; even other sailors, soldiers, and marines carried away by inter-ship or inter-service rivalries. If one passed all these obstacles, the insidious John Barleycorn might still claim victory over the hapless sailor who was charged with AOL plus drunkenness, improper uniform, resisting arrest, or disorderly conduct.

Another group of men accosted service personnel in their own fashion: the homosexuals. Occasionally the tables were turned and the servicemen preyed on them. There was no such thing as "gay rights" in 1941. Homosexuality was considered a loathsome disease, a failure of masculinity and morality. A few navy men thought it perfectly proper to assault "queers" who had propositioned them and take their money and other valuables. The battered gay had little recourse, fearing that publicity would bring police harassment as well as recriminations from people at work or in his social life.

Since it was common knowledge that the statue of King Kamehameha was the usual homosexual rendezvous and the Iolani Palace grounds a favorite place for trysts, the Honolulu police occasionally would arrest a serviceman caught attacking a gay there and turn him over to the shore patrol or the military police.

Prejudice against homosexuals ran deep among all navy rates and ranks. The offending sailor usually received a summary court-martial for "malicious conversion" rather than for the more serious charges of assault and robbery. A conviction was by no means assured, as it was when more conventional charges were pressed. By way of contrast, one of the ship's second-class petty officers was booted out of the navy on a bad conduct discharge that same summer for stealing from his shipmates.

As the summer wore on, it became apparent that the navy's previously high enlistment standards were being relaxed. There was a noticeable decline in the quality of the recruits reporting for shipboard training, a decline both physical and mental. The draft was now in full operation and many young men were choosing the navy as the lesser of evils. A number of them had no business being in a battleship (or any other ship): they were a menace to themselves and to their shipmates. The weeding-out process went on ruthlessly, ranging in degree of severity from transfers to other ships to long sentences at hard labor in navy prisons.

With increasing frequency, the forty-eight-starred Union Jack was broken from the starboard yardarm, grim signal that a general court-martial was being convened with Captain Bunkley serving as president. As the flag was run up and two-blocked, sailors would shake their heads and intone with wry expressions, "The poor bastard!" The question was not whether the accused would be convicted; it was only whether he would serve time at Portsmouth before receiving a dishonorable discharge.

All these convenings added to the captain's growing reputation for ferocious discipline. The sycophants among us cringed and fawned. The all-nav types became ever more attentive to minute regulations. The prudent ones, like myself, grew more cautious, the sea lawyers more profane. And the offenders—those who would not or could not obey the rules—increased in numbers and hostility. The steel door to the brig swung to and fro with metronomic regularity all that summer and fall.

Many crewmen thought that the captain was merciless only with enlisted men. I had heard reports from the bridge and flag radio to the contrary. He treated the junior officers, particularly the reserves, with equal severity and more than one ensign, lieutenant (junior grade), and even full lieutenant were relieved from watch or suspended from duty for minor infractions.

A few stalwarts of the radio gang were not cowed or impressed. Chief Reeves exercised his benevolent autocracy exactly as before, his cigars just as foul as before.* (To those who incurred his displeasure and received peremptory judgment, the imperious motions of his cigar must have resembled the waving of a royal scepter.) A. Q. Beal pursued women and Vat-69 scotch ashore and wrote romantic poetry on board, just as before. And M. G. Johnson talked about the bars and the females of Seattle and San Francisco, Long Beach and Fairbanks, Juneau and Panama City, San Juan and Havana, and dispensed his roughhewn but insightful commentary just as before.

"I've seen a lot of skippers like Bunkley," he said one evening. "They're mean and rough because they're bucking for admiral."

As customary when in Pearl Harbor and not on liberty, we had rolled our mattresses and carried them to the crown of number two turret. There, in the fresh, clean breezes from the Koolaus, we were free from the sweaty confines of our compartment and more than half a ship length away from the oppressive atmosphere of the quarterdeck. Although we shared the nearly flat turret top with a number of other crewmen, we could smoke and talk far into the night if we kept our voices low.

Johnson and I were sitting up, using our mattresses as cushions against a half-foot thickness of armor plate. Now he poked me in the shoulder with an extended forefinger in a characteristic gesture.

"Look at it this way. The sooner he does, the sooner he'll be off this bucket."

Suddenly his expression changed. "There's one little problem, Ted. I don't think he's going to make his stars. He's an old man, and he's running out of time. There's gonna be a war soon, and the navy will weed out all the lousy COs and send them back to shore duty, where they belong. Then they'll bring up some real salty skippers who don't give a shit about personnel inspections and writing reports and firing salutes and running general court-martials—but know how to fight a ship."

Johnson looked across the harbor toward the lights of the navy yard and the sub base, a somber expression on his face.

"The trouble is, Ted, there are a lot of peacetime misfits and they're gonna get a lot of guys killed before the navy wises up."

He lit a cigarette and forced a carefree smile. "Hey, you heard the scuttlebutt? They're rotating the battlewagons back to the States for R & R. Just wait till we hit Market Street, buddy!"

As I came up from main radio one midday in early summer, I noticed a sailor waiting near the hatch coaming. He gave me a tentative, half-shy

---

*As a first-class yeoman, David C. Moser, CY, USN (ret), served with Reeves in the *California* in 1934–38. In a letter of 7 November 1980, he writes: "We used to try to figure the miles of cigars he smoked per year—6 inches to a stogey, one stogey per hour, so many miles per year."

smile, and I returned it. Such greetings were not common outside one's own division in the *California*, where a hard, wary countenance seemed part of the uniform of the day.

"Pardon me," he said. "May I talk to you for a minute?"

I looked at him more closely. He was my size, about five feet nine inches tall, with light-brown hair and a very fair complexion. A bobby-soxed teenager of that time probably would have pronounced him "cute." He could not have been in the navy more than a few months, judging from his still-strong southern accent.

"My name is Tom Gilbert," he said. I introduced myself, and we shook hands. Handshakes were not common among enlisted men either, but it seemed the thing to do with this polite young Southerner.

"You're a radioman third, aren't you?"

I nodded, a little surprised he knew. In some of the divisions, chevrons were stenciled on the T-shirts to indicate petty-officer ratings, but Reeves did not approve of the practice. In the radio gang, only one man had any real authority.

Gilbert told me he was a seaman second in one of the deck divisions but wanted very much to become a striker in the radio gang. So did many others. The members of C-D Division were among the elite of the enlisted men in the battleship navy, along with the quartermasters of N Division and the fire controlmen of F Division. The obvious reply was that the radio gang didn't need any more strikers and there wasn't a prayer of getting accepted. But something about Tom Gilbert, perhaps his good manners, perhaps in part a subtle flattery in his deferential approach, made me decide to talk to him. I suggested we meet at the gedunk stand after chow.

Over coffee I explained the situation. Many highly skilled radiomen from the communication reserve were joining the fleet. The radio schools at the training stations were turning out an increasing number of well-qualified strikers, all of whom had had four months of instruction. Chief Reeves was very tough. He would ask a lot of questions about educational background, typing ability, knowledge of the code, and so forth.

Gilbert's major debit was that he couldn't tell a *dit dah* from a *dah dit*. On the credit side, he had a high-school diploma and knew the rudiments of the touch-typing system. He was neat, clean, and military looking. He was from Alabama: "Yes, sir" and "No, sir" came easily to him. He stood straight and had a fearless honesty that appealed to me. After we talked for a few minutes, I made a quick decision.

"Okay, Tom," I said. "I'll talk to the chief."

Although he was a serious young man, a sudden broad smile illuminated his face. Jumping to his feet, he shook my hand again. "Thank you, Ted!" he said.

It was Reeves's custom to take a number of turns on the forecastle deck after supper, where he was available to any member of the radio gang who needed advice, counsel, or intercession with the division officers. As always, several radiomen were ahead of me, some with problems and some only currying favor. When the chief was free, I walked along with him as he puffed on his cigar.

"What's on your mind, Mason?"

"Chief, I think I've found a seaman second who will make a good striker."

"I have plenty of strikers, Mason."

I remembered my first weeks and months in the radio gang. Had he chosen to, Reeves could have made my life a purgatory. Instead, I had received treatment that was eminently fair and tolerant of my shortcomings as an operator. I felt sure my attitude had favorably impressed the chief, just as Tom Gilbert's attitude had impressed me. I told Reeves about him, not omitting his ignorance concerning Morse code.

"So what makes you think he's striker material?"

"Well, chief, he's bright, he's clean-cut, and he really wants to be in the radio gang. Also," I added with a sly laugh, "he's got good old-fashioned Southern manners. He'll probably call you sir."

The chief stopped short. He removed the cigar from his mouth and regarded me narrowly. Then a twinkle appeared in his eyes. "All right, Mason," he said. "Send him around."

I sent Tom Gilbert around and, against all odds, he was accepted into the C-D Division. He told me about it a few days later, his ordinarily impassive face animated. I was very glad I had been able to help him. How could I have known then that I had just co-signed his death warrant?

Thereafter, I did not see a great deal of Tom Gilbert. The enlisted ranks had their own class system, and petty officers did not ordinarily become close friends with strikers. His move to the forward radiomen's compartment, where I did not regularly go, put additional distance between us. I would see him in the comm office, when he had messenger duty or was diligently tapping away on the practice oscillator. And we would always exchange polite greetings.

"How are you, Tommy? Everything okay?"

"Oh yes, Ted. Everything's just four-oh," he would reply in his liquid Alabama accent.

Since only the most advanced strikers were assigned to radio one and two and the bridge during general quarters, the rest of them were sent to the powder rooms and the ammunition-handling rooms, far below the waterline, where they were hostages to the fortunes of the ship. Tommy Gilbert's battle station was on the first platform deck in one of the handling rooms for the forward turrets. On 7 December a Japanese torpedo hit opposite these spaces; they flooded because of many open manhole covers in the voids of the torpedo defensive system. Tommy never came out alive.

I thought of him many times afterward, looking for guilt, and had no trouble finding it. The two questions I repeatedly asked myself were: Had I been playing big shot in getting him into the radio gang? If I had left well enough alone, would he have survived? The answer to the first question is highly subjective. There is no answer to the second.

His death made me consider, again and again, the burden of account-ability I bore my country, my navy, and my conscience as a responsible petty officer in wartime. I swore to do everything in my power so that not one man would ever be lost through dereliction of duty or temporary failure. And all through the terrible campaigns of the Solomon Islands and Bougainville and Pelileu and the invasion of the Philippines, not one man in my ship was.

But this does not make up for the loss of Tommy Gilbert.

# CHAPTER 10

# "Strength Through Love" in the States

After six months in Hawaiian waters, the good news we had long awaited passed through the radio gang at slightly less than the speed of light.

"The chief just took his dress blues to the tailor shop! Stateside, by God!"

Very soon we were receiving other tip-offs about our imminent departure. A yard oiler came alongside to port and we began defueling ship. (While a battleship normally fueled to capacity before a long sea voyage, we needed much less than that to reach California.) With only the *Neosho* and a couple of other oilers making the run from the West Coast, Pearl Harbor was always short of fuel oil.

Some two dozen men who had been on temporary duty in the sub tender *Argonne* (AG 31), or at Camp Andrews, or with the permanent shore patrol in Honolulu began reporting aboard. From the *Maryland*, *West Virginia*, *Tennessee*, *Saratoga*, *Raleigh*, and *Dewey* came parties of men with bags and hammocks for transportation to California. They were sent to the Z Division berthing spaces, abaft sick bay on the second deck, where the drafts of boots normally received their indoctrination into life in a battleship.

Prior to departure, one dreaded social obligation had to be met: the monthly "short-arm inspection." After their exile in Honolulu, it was necessary that the virile crewmen of the *California* be pure in body, if not in mind, before sampling the temptations of the state for which their ship was named.

With my Christian Science background, I naturally was suspicious and a little fearful of doctors. The few I had seen appeared only when someone had passed on, or was about to. At boot camp, I had quickly learned that the navy was not going to change its routines to accommodate esoteric religious beliefs. The business of the navy, after all, was to keep its men healthy enough to die well in combat, should that prove necessary; and no exceptions were made to the stipulated inoculations, inspections, and treatments. Thus forcibly exposed to the navy's health-care system, I gradually came to see that it was rather admirable.

My only real physical weakness was a susceptibility to colds. In the confined quarters of a battleship, and notwithstanding the highest standards of cleanliness of any navy in the world, these were an ever-present

The author and his close friend Bill Fisher (*left*) have something to laugh about in early October 1941. The *California* is going stateside on a "strength through love" cruise!

hazard, particularly in rainy Bremerton. At first, I suffered mine in silence. But discomfort at last overcame my reservations, and I decided to seek relief at sick bay.

Sick call was held around 0900 daily. A follow-up call in mid-afternoon was devoted mostly to changing bandages, administering medications, and general checkup of those on the "binnacle list." These men were excused from duty but slept and messed in their regular spaces. They were checked daily, with the object of returning them to duty as soon as possible. The official sick list, on the other hand, comprised those men who were actually confined to sick bay. To get on this list, one had to arrange to lose part of a finger on a loading machine, fall headlong down a ladder and break a bone or two, or encounter seven Kanakas in an unlighted alley of a Honolulu slum.

For the morning call, however, it was necessary only to be excused by one's leading petty officer and fall into the sick line. Sick bay occupied the entire athwartships section of the second deck between number one and number two turret barbettes. Except for the piping and conduits suspended from the overhead and the circular ports, it looked (and smelled) like a small hospital. On the starboard side, it was equipped with dressing stations, a laboratory, and a dental office. To port were an examination room, a six-bunk isolation ward, and the compact operating room. The latter was tiled to waist height and painted beige above that, with a battery of shatterproof lamps suspended over the operating table.

Most of the forward transverse bulkhead was occupied by some two dozen luxurious berths, two-high and fat with Simmons innerspring mattresses. We crewmen cast many a covetous glance at these accommodations, so luxurious by comparison with our own pallets. Our only consolation was that those permitted to sleep there were too sick to be able to enjoy it.

Sick call formed along the starboard passageway of the Z Division compartment. The line moved with surprising speed: it was not long before you found yourself at the main dressing station, which was presided over by a first-class pharmacist's mate in spotless whites, assisted by a hospital apprentice or striker. Another corpsman would be on duty at the nearby dressing table. Usually no doctors were in sight; they showed up only after sick call was over.

The first class, always a career man with two or three hitches, would favor you with a cool, clinical look. Sometimes he would ask, in clipped tones, "What is your complaint?"

"Doc, I've got a goddam cold and sore throat. Can't hardly talk!" (This in a hoarse voice, usually accompanied by a racking cough or two.)

Wordlessly, the first class would reach for a tongue depressor and peer down your throat. Flipping the thin wooden stick into a wastebasket, he would say to his apprentice, "Jones, take this man's temperature." A thermometer was popped into your mouth, and you stood aside as the line moved along.

When the glass tube was removed and shown to the first class, a brief inclination of his head brought a conical paper cup filled to the brim with a viscous, dark-brown concoction called Brown Mixture and two "horse pills" (not aspirin but a special cold formula prepared by one of the leading pharmaceutical companies). As soon as you had choked these down, you were dismissed as cured. Should the treatment prove temporary, you could always return for another.

If the pharmacist's mate of the watch detected more serious symptoms, the sailor was turned in to sick bay to await the arrival of one of the doctors. The system seemed to work very well. Those hard-bitten pharmacist's mates were the paramedics of their day, and had the benefit of more thorough training. The big difference, Old Navy corpsmen like William A. Lynch have told me, is that the typical first- or second-class pharmacist's mate of 1941 had ten or fifteen years of navy service, which included independent duty on small ships that did not rate a doctor. As a result, he acquired a broad if nonacademic knowledge that enabled him to sort out complaints from symptoms and make intuitive diagnoses that were surprisingly accurate.

Since only a small percentage of the Honolulu prostitutes—those professionals who worked in the GI houses—could be inspected for venereal disease weekly, the navy had other ways to guard against invasion by virulent microbes from the bars and byways of the city. Once a month, each division was summoned to sick bay at a specific hour. Awaiting the members in Z Division compartment were one of the ship's three medical officers, seated on a stool, and a pharmacist's mate, standing beside him with a flashlight in one hand and a supply of tongue depressors in the other.

As you came abreast of the medical officer, you had to drop your pants and underpants and raise your T-shirt so you were exposed from navel to knees. The corpsman targeted his flashlight on that seemingly insignificant part of your anatomy. Selecting a tongue depressor, the doctor flipped your private parts back and forth, up and down, looking for infestation by crab lice (*Phthirius pubis* to the MDs, but "a dose of crabs" to the enlisted men).

"All right," he would say. "Skin it back."

Skin it back you did.

"Now squeeze it out."

Squeeze it out you did. And unfortunate was the man who discovered a glistening, pearly drop at the apex when this exercise had been completed. He was hustled into sick bay for immediate confirmation of a gonorrheal infection and automatic restriction to the ship until he was cured. Which might take a long time. In 1941 the miracle drug penicillin was not yet available; the latest remedy was sulfanilamide, which was slow and uncertain at best. For the luckless few who contracted syphilis (variously called syph, Old Joe, and the Bug), the prognosis was even more unfavorable. Since it could not be cured but only arrested after a long period of treatment, the victim might be confined to the ship for up to a year.

Among some well-traveled sailors, it was *de rigueur* to scoff at the seriousness of gonorrhea (but not syphilis).

"The clap?" they would ask rhetorically. "Why, it's no worse than a bad cold. Hell, I've had it seven times."

But I knew this was mere whistling past the graveyard. If the navy took VD seriously, I was well advised to do likewise.

Just abaft sick bay, port side, was a prophylactic station. If you had been exposed, all you had to do was sign in on your return from liberty and take a "pro" and you were in the clear so far as navy punishment was concerned. And probably clear of venereal infection, as well—if you got there soon enough.

The procedure, however, was certain to remove all illusions about the romance of sex. First you washed with soap and water; next a purple liquid (containing argyrol) was injected up the urethra and held there for five minutes; then a whitish paste, called Calomel ointment, had to be liberally smeared over all exposed parts and left there overnight. Finally you had to endure good-natured ribbing from shipmates the next morning when washing the gooey mess off in the shower. Still, most men understood how sensible it all was. The navy was protecting not only its battle efficiency but also the welfare of its people.

As if the short-arm inspections were not humiliation enough, some of the participants found themselves reluctant recipients of minor surgery. The medical officer who usually performed the inspections had a rather bizarre quirk. He liked to perform circumcisions. And here was an unlimited supply of potential customers.

It was not automatic in those days to perform this minor sanitary and ritualistic operation on male babies. At least half the navy came from farms and small towns where medical practices were unsophisticated.

As the *California*'s doctor flipped organs with his tongue depressor, he would announce in a gruff voice to an unsuspecting seaman, "You need a circumcision."

The sailor would stare at this omnipotent figure—not only a doctor but an officer as well—in fear and wonder. More often than not, he didn't even know what the word meant. If he lacked the will or the confidence to voice strong objections, he would shortly find himself in the ship's surgery.

As the number of seamen with bandaged members mounted, the medical officer's notoriety spread throughout the ship. A bawdy couplet composed by one of the rude, untutored poets among the ship's company speeded the word along. Properly forewarned, newcomers were able to resist the diagnoses and surgical procedures of the doctor who was "queer for circumcisions." By the time of our departure for the West Coast, there were few volunteers for his services.

On Sunday, 12 October, I was able to celebrate one year on board the *California* by waving to spectators as we passed under the Golden Gate Bridge. And the best was yet to come: Fisher and I were scheduled to depart the next day on glorious leave and would rejoin the ship in San Pedro Bay.

Before our leave papers could be signed, we were impatient passengers on a short trip to Hunter's Point, where the *California* entered dry dock for an emergency repair, this time to the auxiliary overboard discharge valve in the after engine room. The din of chipping hammers soon reverberated throughout the ship as deck hands went over the side to purge the marine growths that had helped slow us to 17.4 knots (well below our rated speed of 21) on a full-power run not long before.

While we were in dry dock, the story of our Nazi radio striker grew even more puzzling. I happened to be topside when the striker appeared on the quarterdeck in dress blues, escorted by Spud Murphy himself. The scowl on Murphy's broad bulldog face was even more forbidding than usual. I got the impression he would have welcomed an opportunity to use his service .45.

As he nudged the blond striker across the brow, and none too gently, two tall, well-groomed civilians in dark, double-breasted suits closed in. They looked like a Hollywood casting of FBI agents; I had no doubt they were the real thing. They put the striker into the back seat of a black government sedan, and that is the last we ever saw of him.

If he was in fact planted in the navy as a German spy, he was obvious to the point of caricature and, consequently, of little value. Later it occurred to me that he might have been a counter-agent, placed by naval intelligence to ferret out potential Nazi sympathizers. If that was his role, it is doubtful that he enjoyed any success. As I have indicated, there were a number of career petty officers who admired the German people for their efficiency, technology, and military prowess. But they considered the Nazis an aberration. In their opinion, Hitler was a ranting psychopath, and his jackbooted, goose-stepping storm troopers were a sort of Praetorian Guard, useful mainly for impressing civilians. The marines, they said, would go through them "like shit through a tin horn."

Night was falling before Fisher and I could pick up our leave papers and flee the ship. My father, driving a long blue 1939 LaSalle, picked us up at the ferry building. Normally a cold and a captious, even somber man, he was delighted to see me.

At the house on State Street, Bayard Bowman got out a bottle of bourbon.

"Since you're old enough to serve your country, son," he said, "you're old enough to have a drink."

After dinner with my father and his second wife Carol, Fisher was impatient to go, and so in a way was I. Johnson's girl friend Eleanor had driven down from Seattle and rented an apartment on Pine Street. She had brought Mickey, the little sailor's girl with the congenital heart condition, and another friend.

I knew my father wanted me to stay, especially since we so rarely saw one another. I also wanted that, but there was too little time in which to bridge so many years. With only a week or so of leave and three weeks in the States, time was a precious commodity, and Fisher and I rationed it with the ruthless selfishness of youth.

We called a cab and hurried to Pine and Powell. The celebration went on most of the night, necessitating frequent trips to the liquor store down the street as more shipmates and girls arrived.

When Johnson held reveille in the gray chill of the San Francisco morning, we should have been hung over, but we were not. There was no time for that sort of weakness. A shave and shower, coffee with Johnson, and breakfast at one of the Log Cabin short-order cafes that had proliferated in the central district during the Depression, and Fisher and I were almost as good as new. For an investment of fifty cents each, we took a cab to the Bay Bridge approach and set out hitchhiking home to Placerville.

In 1941 it was easy to get a ride if you were in uniform. Civilians, aware that sailors at liberty were always short of both time and money, were anxious to help. In exchange, they hoped to learn what the armed forces were doing. Once past the formalities, such as their own military service during the Great War and the nephew or in-law who had just been drafted, they would ask what ship we were on. When they heard we had just come from the remote outpost of Pearl Harbor, their typical question would be, "Do you think the slant-eyes are going to give us any trouble out there?"

Somewhat pompously, and in our infinite ignorance, we would assure them that the Japs wouldn't dare; and should they be so stupid, we would promptly kick the hell out of them. That sally would draw appreciative chuckles; they were relieved to know that "our boys" were guarding the ramparts.

By rapid stages we passed Berkeley, where I had been born, and Vallejo, where my ship had been commissioned at about the same time, and crossed the M Street Bridge into Sacramento, whose Capitol Park was to house one of the few relics of the *California*—her tarnished ship's bell. Beyond the city, the highway narrowed and twisted like a snake as it began to climb the lower foothills of the Sierra Nevada. In the distance, the sun glinted off snow-covered peaks and towering escarpments of pale-gray granite.

We covered the 150 miles from San Francisco in about four hours. By early afternoon we were alighting in front of a busy little market and service station in the Placerville outpost of Diamond Springs, once a thriving gold-rush town, now a hamlet dependent upon its lumber mill. The market was owned by Fisher's elder brother, Ted. Very soon we were grandly tooling Ted's Packard down county roads toward Missouri Flat, itself once a Mother Lode "diggins." We passed the one-room grammar school where I had completed eight grades in seven years and the community hall I had helped to build, and turned onto the rutted lane that led to my foster-parents' home.

"My folks will be expecting to see you tonight," Fisher reminded me as he drove away.

The Masons owned a tin-roofed cottage that lacked running water and electricity. Its charms were all external, for it nestled under a dense canopy of evergreens on a terrace above Indian Creek, with a view across meadows and grazing lands and wooded hills to a lonesome peak in the distance. Behind me the hill climbed steeply in groves of live oak and digger pine and ponderosa, interspersed with thickets of toyon, bearing bright-red berries, and wine-dark manzanita.

With Fisher gone, it was very quiet there. The silence was broken only by the tuneless twitterings of finches and sparrows and the mocking, strident call of the blue jay close at hand. Then I heard a cow bell; the occasional faint rumble of a car on the county road; and the dull *pop-pop* of a .22 rifle in the distance as someone hunted jackrabbits or ground squirrels, the way I and my dogs had once done. I was a long way from Pearl Harbor, the bars and bagnios of Honolulu, and the pitiless cold face of the commanding officer of the *California*. The drumming attack of an acorn woodpecker on a tall white oak across the creek reminded me of that.

After I greeted the family, my Uncle Horace invited me outside, ostensibly to try out his new Colt Woodsman pistol but actually to share "a couple of snorts" of whiskey before supper. The younger bachelor brother of Dad Mason, he worked as a night watchman to support the family. (My monthly allotment check of fifteen dollars provided a few luxuries, such as steaks and batteries for the radio.) Since Mrs. Mason disapproved of liquor, Horace was not allowed to drink in the house; he kept a bottle hidden under his mattress. During my adolescence, he had often been the only adult I could share any confidences with. A few more were shared as we passed the bottle back and forth.

Over dinner I gave my foster-parents, to whom I never felt very close, a carefully expurgated account of my adventures, with emphasis on the facilities of the various YMCAs and the number of Christian Science churches I had attended. For the tenth or fifteenth time in my memory, Dad Mason talked about the relative who had been with Admiral Dewey at Manila. He had put in his thirty years and retired as a chief or warrant officer. All during the Depression, Dad used to recall enviously, his relative had drawn a retirement pay of seven dollars a day, a princely sum in those days.

The *California* lies at anchor in Commencement Bay at Tacoma, Washington, during a 1936 good-will cruise. The immaculate condition of the forecastle, forward turrets, superstructure, and foremast are the result of battleship navy "spit and polish." (Photo courtesy of David C. Moser.)

I understood the point Dad always made: the navy was an excellent career, with lifetime retirement pay waiting at the end of two or three decades. But I cared nothing about a goal so far in the future. I had a better reason for being in the navy.

During the 1930s, it was the custom of the navy to send one or two capital ships to every Pacific Coast port, major and minor, on national holidays as a goodwill gesture. One Fourth of July, when I was twelve, I visited the boardwalk amusement area in Santa Cruz with foster relatives. Offshore, a battleship was anchored. She was the largest and most impressive thing I had ever seen afloat; her lines seemed to radiate power and menace. She might even have been the *California*.

Standing alone near the Santa Cruz pier was a sailor from that ship. He was young and lean and clean, wearing a perfectly tailored dress-blue uniform with the single red chevron of a third-class petty officer. He surveyed the mob of civilians nonchalantly, looking worldly and quite removed from their mundane concerns. He was a sailor in the United States Navy! Staring at him but not daring to speak, I thought that he was an even more romantic figure than the heroes of Zane Grey's *Riders of the Purple Sage* or Sabatini's *Scaramouche*. The image I received of that sailor was fleeting, but it engraved itself indelibly in my memory. That was the moment I decided to join the navy one day.

Now, I, too, was a third-class petty officer: I had become that lean, aloof figure I had so admired at Santa Cruz.

With supper over, the gasoline lanterns fired up, and Gabriel Heatter pontificating about the news of the day, I had to ration my time again. I borrowed the car—the 1934 Ford V-8 sedan I had often driven—and went to the Fisher's, where I would spend most of my time thereafter.

They lived in an old wood-frame company house at the Diamond Match lumber mill in Diamond Springs. Lying in the shadow of a large planing mill, the house seldom got any sun and had not been painted for many years. But inside there was noise and gaiety and relaxed acceptance. A florid, stout woman who had once been blonde, Mom busied herself preparing huge meals and fussing over us. Pop, a time- and work-worn version of Bill, chain-smoked cigarettes (he called them *see-gar-'eats*), plied us with beer, and haw-hawed at our sea stories. He had not forgotten what it was like to be young and restless: there was no curfew, and the family car was always available to us.

We were on the go constantly. The days and nights passed in a blur of motion, and I had to check the calendar daily to remind myself how little time remained. But we remembered our professional obligations and spent a few pleasant hours with the local ham operator, perched on a hilltop above the Highway 50 bridge across Weber Creek. To our surprise and gratification, we had already surpassed him in operating skills, although he was twice our age and had much more experience. Like many hams, he was more interested in upgrading his rig than in the proficient operation of it.

The last full day we could spend at home was a Saturday. The place to go that evening was a country dance at Green Valley, some fifteen miles from Placerville along another of the narrow, circuitous county roads. This being a special occasion, Fisher again borrowed his brother's Packard.

The once-white community hall was set well back from the road in a grove of pine and oak trees. Admission was one dollar; your hand was stamped with an invisible ink that only showed up under fluorescent light. Since the cars served as portable bars, among other things, being able to come and go at will was most important to the success of the evening.

At the far end of the shoebox-shaped hall, the stage was occupied by a home-grown band that almost made up in enthusiasm what it lacked in talent. It comprised a piano, saxophone, guitar, fiddle, and drums. The music ranged from old waltzes and fox-trots through cowboy songs to a few current hits. The group's output was tinnily amplified with an ancient and temperamental public-address system. It was assumed, even then, that everyone had a hearing defect.

The worn hardwood of the big dance floor was made suitably slippery, at least in spots, by the liberal application of white powder. Along both sides, rickety folding chairs held a cross section of the community: everyone, in fact, but the ill, the infirm, the "holy rollers," and those too poor to pay the admission fee.

Girls were everywhere: gathered in small giggling groups; dancing with the boys who were bold enough to ask or with their girl friends; walking a little too rapidly and all too consciously toward the refreshment stand or the restroom. And one of the loveliest was a girl I'll call Lola O'Dell. Her older sister Maggie, now married, had been in my high-school class. When I had last seen Lola, she was a thirteen-year-old beanpole with a chest as flat as a boy's and knobby knees. In the time between, the truly magical transformation from puberty to womanly sexuality had taken place—a metamorphosis almost as dramatic as that from a gray larva to a monarch butterfly—and I couldn't keep my eyes off her as she danced with a pimply high schooler.

I nudged Fisher. "That's Maggie O'Dell's little sister. Over there by the bandstand. Jee-zuzz!"

"Yeah," my buddy responded. "Not bad. A little skinny for my taste maybe." His roving eye caught another face and he actually smiled. "Hey," he said, pointing toward a freckled, red-haired girl of more substantial figure. "There's Betty Farrell. See you later, Mason."

I went over and talked to Lola. Up close she was dazzling, with perfect teeth, a small nose, shoulder-length blonde hair, and a pink-and-white complexion. She remembered me well: her sister had often spoken of me.

When we danced, she moved close. She was so lithe and graceful, steering me so subtly through the steps of a fast fox-trot, that I could almost fancy myself an accomplished dancer. Wondering at my good fortune, I asked her why the football players and the future farmers weren't monopolizing her.

She wrinkled her nose. Her eyes, a shade of deep and smoky gray, were wide and candid as she looked up at me. "Oh, they're so young," she said. "They don't know anything."

I held her a little closer and silently thanked my shipmate Johnson for the months of tutoring in this very subject. From there, things progressed rapidly. I invited her to the car for a drink, and she smiled and looked up at me with those smoldering Irish eyes and accepted.

We sat on gray mohair upholstery in the back seat of the Packard and I prepared a Cokeball, country style. A bottle of Coca-Cola was opened and half of its contents dumped out. Blended whiskey, purchased through the offices of an older acquaintance at a Placerville liquor store, was poured in to replace the loss. Using the cap as a stopper, the bottle was inverted to mix the ingredients. Lola and I passed the bottle back and forth, making faces as the warmish, fiery stuff went down.

Soon the unheated car was cozy despite the near-freezing temperature. The shouts and laughter from neighboring cars receded, and the stars seemed as brilliant as diamonds against black velvet. Even the cold pyramidal shapes of the ponderosa pines just outside the windows softened and sighed benevolently, as if in a spring breeze.

After necking, which progressed inevitably to petting, she pushed me away gently and sat up. I found the Cokeball, propped precariously against an armrest, and passed it to her. She shook her hair, pulled down her sweater, took a sip, and shuddered.

"Would you like to sleep with me, Ted?" she asked. It is not true that today's woman invented the approach direct.

"Yes, Lola," I said enthusiastically. "I sure would! But where?"

"Well, not at my mother's place. My God! We'll go to my sister's. She likes you."

The prospect was enticing. But there was a problem. Our leave expired at 2400 Sunday. The *California* was anchored in San Pedro Bay, more than 450 miles away. It was already early Sunday morning. If I stayed with Lola, I would surely be AOL.

I thought about the dour and implacable Captain Bunkley. I thought about my thin, pristine bunk in compartment B-511-L. I thought about the claustrophobic radio shack on the first platform deck. I thought about Lola without clothes. For the first time in my navy service, I faltered in my clear duty.

Ahead of me, at the very least, were more regimentation, more spit and polish, more battle exercises off the Hawaiian Islands. Perhaps war with Japan. Death, even, I thought dramatically. And here was Lola O'Dell, young and beautiful and mine for the taking. So I would go to the brig for a few days. Better men than I—men like my great friend Johnson—had done their solitary confinement on bread and water and had survived.

"To hell with it. I'll stay."

We were sealing this Faustian bargain with a long horizontal kiss when I heard a rapping on the right rear window. I ignored it. But it only became louder.

"Shove off, whoever you are!" I snarled.

Now the door opened and Fisher's stern face filled the void.

"All right, Mason," he said. "Reveille. Grab for your socks. We're heading for Diamond Springs to meet Lyle. He's driving us to Sacramento." Behind him in the dim starlight, several of our old high-school friends were grinning idiotically.

"You guys shove off," I said, reaching for the bottle. "I'm staying here with Lola."

"You can't stay here, Mason," Fisher explained in the tone one reserves for the young and stupid. "Liberty expires at 2400. In Long Beach, at the fleet landing. So say goodnight to your girl friend and let's go."

"Frig the fleet landing," I said with feeling. "Frig 2400. Frig Captain Bligh-Bunkley. Frig Captain Bligh-Bunkley in the rigging. I'm staying with Lola."

Fisher decided to try diplomacy. He turned to Lola.

"Look, honey," he said. "This crazy bastard Mason is drunk. He thinks he's in love. But he's gotta be back at the ship this evening by 2400. You don't want him to go to the brig, do you?"

Lola shrugged. "But he is in love," she said, as if that made everything right. She pressed herself against me. "You are in love with me, aren't you, Ted?"

"You're damn right I am," I said. We fell into another embrace. "Shove off, Fisher," I repeated when we came up for air. "To hell with that goddam—"

My friend broke in with an expletive. Then, dimly, I realized he was conferring with our friends. "All right, Mason," he said in a conciliatory voice. "But you gotta let me have the car."

"That's four-oh with me." I gathered up my peacoat and the whiskey and Coke bottles, and Lola and I emerged. She took one look and started pulling me away. "Let's go, Ted. Over here."

Too late, I realized I had been outmaneuvered. Three pairs of strong arms seized me and pulled me away from Lola. I and my belongings were flung back into the Packard. Lola broke away from Fisher's grasp and fell into the back seat on top of me. We clung to each other desperately but in vain. One traitorous friend pried Lola off while the other two blocked my egress. Fisher jumped into the front and started up the car.

"Ted!" Lola was screaming. "Stay with me! Don't let them do this to us! Ted!"

But the car was backing and turning and there was no escape. "Lola!" I shouted out the window. "I'll be back. Wait for me!"

Even as I spoke the words, I knew they would be of no avail. Soon, she would be claimed by another: a lubberly soldier or, even worse, some flat-footed civilian.

As we shot past, I caught a glimpse of Lola in the glare of the headlights. She was bent forward, crying and pulling at her hair.

The Packard turned onto the Green Valley Road and sped toward Diamond Springs, pitching and swaying. In the back seat, I was groaning and cursing those who had saved me from myself.

For on that October night in 1941, I was only one young sailor who wanted one young woman. Of course, I never saw her again.

We returned to the ship late Sunday evening with a couple of hours to spare, having used various modes of transportation: auto to Sacramento, Greyhound Bus to Los Angeles, Big Red Car to Long Beach, and taxicab to the Pico Avenue Landing.

As the motor launch traversed the Outer Harbor, I could see the pale silhouette of the *California* looming over the darker gray of the breakwater. We had left the civilian world at the landing and were about to reenter that much sterner and more unforgiving one of the battleship navy. Now Placerville and the Masons and the girl seemed far away. With a pang of regret, I realized I didn't really have a family. My own father and mother were near strangers, my foster-parents more so now than they had ever been. Now my family was made up of friends and shipmates, like Fisher and Johnson and our strict but fair chief, and the *California* was my home.

I put my arm around Fisher. "Hey, buddy. Thanks for dragging me away from Lola. I appreciate it."

Fisher looked embarrassed. "Aw, forget it, Ted," he mumbled. "You woulda done the same for me." Quickly, he extricated himself. "Not that I'd ever do anything so stupid!"

But he had used my first name, and I knew what he meant. "Of course you wouldn't, Bill," I agreed. And that was the end of what could have been a most costly foul-up on my part.

"Where the hell you guys been? Don't you know there's an alert on?"

We didn't have to pretend ignorance: we hadn't even read a newspaper during our short but memorable leave. Now we were told that, on 17

October, the crew had been mustered and informed that relations with Japan were at a possibly critical phase. All leaves had been cancelled. A number of officers and men had been called back to the ship (some from as far away as the East Coast). Henceforth, everyone going on liberty had to remain within two hours' recall time.

All this brouhaha appeared to have resulted from the fall of the Japanese cabinet and the seizure of the premiership by General Hideki Tojo, leader of the war clique in the military establishment. Judging by the photos and newsreels of Tojo we had seen, it was hard to take him seriously. He looked and acted like a Gilbert and Sullivan parody of an Oriental military man, bullet-headed, bespectacled, bellicose. Remembering the phony Honolulu–Pearl Harbor alert of the past April, the consensus in the petty officers' head was that Kimmel had pulled "another boot-camp drill," without regard for the personal hardships it would cause. We were in the States for R & R, for crissakes! (More picturesquely, some of the junior officers termed it a "strength through love" cruise, in mocking reference to the highly publicized "strength through joy" cruises of the Hitler Youth down the Rhine.)

The two-hour limit didn't work any great hardship on me. Johnson's friend Eleanor and her entourage had moved to an apartment on Cherry Avenue, and that is where I spent many of my liberty hours. The apartment was only a few blocks from Bluff Park, which overlooked San Pedro Bay, and was less than fifteen minutes by cab from the fleet landing. Since our last visit to Long Beach, I had learned a good deal—there would be no street-crossing duty with the little old ladies this time. If I were with Johnson or Red Goff, I could even buy a drink in certain navy-oriented bars that didn't check IDs carefully.

The same thing was true in Los Angeles, where I was served in establishments that ran the gamut from the art-deco squalor of the Waldorf Cellars on Main Street to the chrome and red-leather and potted-palm chic of the Florentine Gardens in glamorous Hollywood. The women, too, were hospitable, especially when we borrowed Goff's 1938 Ford two-door (in mint condition because it had spent most of its three-year life in storage in Long Beach) and took them to the Gardens or Earl Carroll's. Since such an evening could cost fifteen or twenty dollars apiece, we ran out of money even before we ran out of time.

At the Cherry Avenue apartment Johnson began to act the surprising role of police petty officer. At 0600 all hands would be awakened by a loud and maddeningly cheerful voice.

"Now up all idlers and prepare to relieve the watch," Johnson would sing out. "There'll be no brig time this time!"

Properly fortified with cups of freshly brewed coffee, we would report on board well before liberty expired at 0800.

But few of the *California* crewmen enjoyed such close and rewarding friendships, and not many were having the glorious time that I was. The mood of the ship was entirely different from what it had been on our first trip to the States. In late 1940 there had been an exuberance about the return to the promised land that was lacking now. This time there was a desperate edge to the celebration, an air of sullen and hostile defiance.

Already disliking the captain for his remorseless discipline, many of the younger sailors put themselves into positions where they were obliged to hate him for the punishments they incurred. These frustrated ones were locked up in the brig or, since the brig was always filled to capacity, were prisoners at large, restricted to the ship while they awaited their turn. As the latter watched the liberty boats cast off, they must have longed for Pearl Harbor, where confinement to ship was a matter of little consequence. Now I understood that curious masochism by which men deliver themselves up to their oppressors. But for Fisher, I would have been one of their rapidly growing numbers.

On my last liberty, Mickey accompanied me to the Pico Avenue Landing. To ward off the damp chill of the Long Beach midnight, she was wearing my peacoat over a gaudily flowered dress. Around her neck was a gold cross some sailor had given her. The thick straight dark hair that was her best feature flowed down over her shoulders. She was not a Bertha Morton or a Mary Jane or a Lola of the country dance, and her eyes were rather bleak, perhaps masking pain and fear, but she had been kind to me, in her fashion. And I could not forget that her time on this earth might be short.

We had a beer at one of the tables outside the civilian cafe. The salt air was laced with the pungent smell of crude oil. The lights of the harbor piers and terminals shone coldly across the water. Liberty boats came and went. The more fortunate sailors had someone to say good-by to, the others drifted past and were gone. Out of the low opaque clouds that cloaked the Outer Harbor came the deep throbbing note of the foghorn.

Mickey smiled at a petty officer seated at a nearby table and remarked idly that she had met him in Seattle. He was very nice, she added. I noticed that he also was exceptionally handsome. I asked if she were going back to Seattle as soon as the *California* shoved off.

"I don't think so," she said. "Our rent isn't up yet. I'd like to stay a few more days."

I knew what she meant, but it didn't matter. Let her fling roses riotously while she could.

A motor launch from the *California* made a landing at one of the floats below us with a ringing of bells and the mushy churning of a backing propeller. It was the last liberty boat until the 0500 milk run.

"Take care of Johnny, Ted. Bring him back to Eleanor."

I said I would. We got up, and I buttoned my peacoat around her. "Keep it to remember me by in those Seattle winters," I said.

I could tell by her eyes that she was moved. She gave me a sisterly kiss.

"You're a good guy, sailor," she said softly.

"And you're a good gal, Mickey," I responded.

We got under way for Pearl Harbor at 0929 on Saturday, 1 November. Between 0015 and 0830, fifteen AOLs had straggled aboard, to be made charges of the new master-at-arms. (Spud Murphy, his job of cleaning out the ship's petty rackets accomplished, had been transferred to well-deserved shore duty.) Twelve men had missed the ship: five seamen, four mess attendants, and three marine privates. The brig was full.

Johnson and I lingered on the quarterdeck. Looking ahead as we neared the breakwater, I could see a bank of dense gray fog. But my friend was looking back at the silhouette of the city. It was then that he said, in a low emotional voice: "God damn it, Ted. There it goes. We're leaving, God damn it, and I'm never gonna see the States again."

Of that lovely and hostile and hungover crew, ninety-eight would never see the States again.

CHAPTER 11

# Reverie on the Midwatch

*A little neglect may breed mischief: for want of a nail the shoe was lost; for want of a shoe the horse was lost; and for want of a horse the rider was lost.*
Benjamin Franklin

"Mason. Hey, Mason. Reveille!"

Dimly, I heard a gruff voice, felt a rough touch on my shoulder.

"You got the mid, Mason. Reveille!"

One of our strikers was doing his duty. I propped myself up groggily on one elbow.

"You awake, Mason?"

"What time is it?"

"It's oh oh thirty. You got the mid!"

"Okay. Okay!"

I had had only two hours of restless sleep, but there was nothing unusual about that. I hit the deck and made a run to the petty officers' head. When I put the toilet kit back in my locker, I got out my billfold and tucked it into a pocket of my white shorts. Why, I didn't know. I certainly couldn't spend any money on the midwatch, and the ship's service store was closed until after church services.

I stumbled down the ladder into main radio. It was 0045 on Sunday morning, 7 December 1941. I had time for a few gulps of coffee and a little conversation to clear the cobwebs before I relieved the watch. The supervisor was John R. Mazeau, a slim, dark-haired second-class radioman in the flag allowance, one of the two "French lover-boys" in the radio gang. Although he was a year and a half older than I was and had been in the navy six months longer, he was baby-faced and looked about seventeen.

My circuit was 2,716 kc, which linked CinCPac with all navy ships present in the Hawaiian area. It was an emergency circuit that usually remained quiet on the midwatch. Tonight was no exception. I had plenty of time to "shoot the breeze" with the other watch standers about our recent cruise to the States and to speculate on when we might return. The consensus was that the Japanese were running a bluff and would pull back at the last moment. Then, perhaps, the battlewagons could leave Wai Momi ("pearl water") to its shark god and return to San Pedro, where they belonged. After the first burst of conversation trailed off into tedium, I filled the time with idle reverie.

I thought about the flying lesson I would have that afternoon at John Rodgers Airport. Just one more five-dollar session practicing takeoffs and landings and I would solo. I still didn't feel quite comfortable when I was aloft. The machine was controlling me, rather than I the machine, just as it had when I was first learning to drive an automobile. Time would remedy that—and time was, I thought, my ally.

*0100. Japanese striking force of six carriers, two battleships, two heavy cruisers, nine destroyers racing south at 26 knots along the meridian of Oahu, longitude 158° west, to launching point 275 miles north of the island.*

I got out my billfold and checked my funds. I still had seven dollars and change from the twenty dollars I had borrowed from Chief Reeves and split with Johnson. That would cover the lesson, with beer money left over. When payday came and I repaid the loan in full, I would be a little short, of course. I had been told the chief was very lenient about granting extensions, but I had no intention of asking for one. That twenty dollars was a symbol of my full membership in the radio gang—and in the battleship navy.

On a vagrant impulse, I removed the tiny photo of Bertha Morton I had carried all these sixteen months in the navy. She laughed back at me, as lovely as ever. I wondered what she was doing now, at this very minute. Hell, I told myself, what could she be doing? California was on zone plus eight time: it was four in the morning there. She was probably pregnant again. I had an image of her voluptuous figure grossly swollen with a 4-F's child. I hastily put the photo back in my billfold.

For the twentieth time, I debated whether I should put in for a transfer to the destroyer navy. Not that it would do any good: Red Goff had been trying for fifteen months to get off "this horseshit flag." Besides, I would have to leave my friends—and that I was most reluctant to do. So long as I had good friends and so long as Chief Reeves was running the radio gang, I was better off staying where I was.

But that might change soon. The scuttlebutt was that the chief had just made warrant radio electrician. Then he would move into that curious limbo of the broken stripe: neither officer nor enlisted man, neither gentleman nor gob. He would inevitably be transferred, probably to CinCPac, and the finest chief radioman in the fleet would be lost to the *California* and to ComBatFor. His place would be taken by the blustering Flag Chief Reinhardt, about whose competence I had my doubts. I hoped the rumor was wrong, or that if true, the chief would turn down warrant as he had reputedly turned down a commission.

If Admiral Pye were half as brilliant as his advancement from Commander Destroyers, Scouting Force all the way to Commander Battle Force in only three and a half years would seem to indicate, he would keep Reeves right where he was. His predecessor as ComBatFor, Admiral Snyder, had held the chief in the highest regard. In the summer of 1940, Reeves's mother had died while the *California* was at sea. Snyder sent the chief to Pearl Harbor in his own plane, the Blue Goose, so he could catch the Pan Am Clipper to San Francisco and a connecting flight to the East Coast to attend the funeral.

The other unsettling prospect, if I continued improving my skills, was that I might get promoted to the flag complement, which in turn might bring an assignment to flag radio. That I would flatly refuse, regardless of the consequences. To me, all the evils of hosting a flag staff were personified in the scowling face of Lieutenant Commander H. O. Hansen, the flag communication officer whom Admiral Pye had brought with him from the *West Virginia* when he fleeted up to ComBatFor the January before.

In appearance and manner, the fleshy Hansen perfectly fitted my conception of a high-octane corporation vice-president, right down to the squinty-eyed, malevolent gaze, the ranting voice, and the fat cigar. He was known and feared throughout the C-D Division as HOH/30, a designation he apparently adopted lest anyone forget he was a graduate of the U.S. Naval Academy, class of 1930.

His specialty was stalking the flag bridge and browbeating radiomen and signalmen. The two second-class radiomen who manned flag radio would often stagger into our living spaces following a watch at sea, pale and shaken.

"Old HOH/30 was on a rampage tonight," they would report. "He was stompin' the deck and climbin' the bulkheads. We couldn't hardly copy through the W-5 interference." (Interference with radio signals was rated on a scale of one to five. W-5 was maximum.)

"How can you take that crap?" I once asked.

One of the flag radiomen looked at me as if I were none too bright. "What the hell do you suggest we do about it, Mason?"

Respect for authority was one thing; allowing oneself to be humiliated by authority was another. "For a start," I said, "I'd get off the flag bridge—and out of the flag complement, if necessary."

The pair stared at me incredulously. "And give up all that extra liberty?" one of them scoffed. "You must be crazy, Mason." Clearly, the prestige and privileges of flag status compensated in their minds for the abusive tirades of the flag communication officer.

For many of the flag personnel, the privileges were so extensive as to elevate them to a near untouchable position. The crews of the admiral's gig and the flag officers' motor boat, for instance, did nothing except operate and maintain them. They stood no watches, were exempt from captain's inspection, mustered on the honor system, and had no general quarters duties except to stand by their craft. They held special ("S") liberty cards that allowed them to go ashore whenever it didn't interfere with their sole job of running the boats between the ship and the officers' club landing. They even lived in the boats. Quite rightly, the envious ship's company thought they had it made.

But Reeves would not tolerate such gross inequities in the labor and status of the flag and ship radiomen. The radio gang was thoroughly integrated; the only special advantages possessed by the flag complement were the "S" card and the esprit de corps that came from knowing that they had been personally chosen by the chief and therefore met his exacting standards. But I did not think an "S" card would be worth the price of enduring a man I considered an arrant bully and tyrant.

Duty in the *California* was getting rougher and more GI all the time. Whether our health cruise (as Kimmel termed the rotation of the battleships to the West Coast) had improved the morale of the crew was arguable. That it had done nothing for the disposition of the captain was certain. On the bulletin boards, the published courts-martial were thumbtacked four or five deep in serried rows—edifying reading for all who might be tempted to err. In front of the master-at-arms shack, dozens of men could be seen mustering several times daily while they awaited their brief hours in court and the inevitable restrictions, fines, and brig time.

Among these unfortunates were the four mess attendants who had missed the ship in Long Beach. After five days of temporary freedom, they had been arrested by the Los Angeles Police Department and turned over to the shore patrol. The *Neosho* provided transportation to Pearl. They were given summary courts-martial on 24 November.

The 1940–41 navy was, of course, totally segregated. The country wanted it that way; the navy liked it that way. The officers' cooks, stewards, and mess attendants were all Negro, Filipino, or "Guam boys." They had their own bunk room, just forward of wardroom country on the third deck, starboard side. It was perhaps 200 feet aft of my own living compartment; it might as well have been 2 leagues.

Their lives revolved around the needs of the officers. The only interaction they had with the crew was at general quarters, when they joined the seamen in the powder and ammunition-handling rooms. Even on liberty, they congregated at their own bars, restaurants, and bordellos.

I seldom saw a member of this group, except at the ship's service store or in the liberty launch, and I don't recall ever talking to one. In a ship as rigidly compartmentalized as the *California*, where each division lived and worked apart from every other division, that did not seem strange to me.

Most of us were not consciously prejudiced. We simply didn't give the matter of racism a thought. Everyone had ample problems of their own without borrowing anyone else's. Besides, we didn't think officers' servants had any special problems. So far as the Filipinos and Guamanians were concerned, they seemed better off in the U.S. Navy than they would have been in their own countries. When they retired, they would be able to live like aristocrats in Manila or Agana.

The Negroes or "coloreds" (the word *black* was not in favor then) were considered equally fortunate to be accepted by the navy, where they had status equivalent to Pullman porters or waiters in private clubs or the better restaurants. Only a few Negroes could belong to their aristocracy of jazz musicians, preachers, and prize fighters; most of them were limited to being boot blacks, house servants, or sharecroppers.

Most of the crew (and, I am sure, the officers too) accepted the proposition that Negroes were inferior in certain ways. To be sure, they were gifted musically and physically: they could sing, dance, run, and box. Jesse Owens had won four gold medals at the Berlin Olympics of 1936; Joe Louis, Detroit's "Brown Bomber," was heavyweight champion of the world. But in the ways that really counted, they were considered simple primitives, hopelessly lacking in mental ability.

Such attitudes were nearly universal, and no one felt obliged to apologize for them. It was deliciously ironic, then, that the four black mess attendants were saved by the very severity of the system that controlled the activities of all enlisted men, of whatever color. The three summary courts-martial boards being overloaded with judgmental duties, there was not time to bring the ship jumpers to trial before the Japanese striking force interfered with the working of navy justice.

In another incident with a less happy consequence, a seaman second (white, naturally) had missed the ship in San Francisco on 15 October. He had been charged with that offense and the resulting AOL of eight hours and thirty minutes before he reported in to the shore patrol. When his summary court was finally published, on 6 December, he was sentenced to the brig for ten days' solitary on bread and water and fined eighteen dollars. He was one of several sailors who did not leave the brig alive. One can speculate that had he been promptly tried and punished, he probably would have been among the survivors. At least, the odds would have been more in his favor. (One can also speculate that I, too, might have been in the brig on 7 December had not Fisher acted as a true friend and shipmate at that country dance near Placerville.)

As if to balance the scales, Fortune bestowed her fickle smile on a number of others. On 2 December, two of our strikers left the ship for the fleet radio school, and R. E. Osborne, a fellow radioman third, was transferred to the auxiliary ship *Sumner* (AG 32). Three days later, a fireman second class was sent to the hospital ship *Solace* for treatment. "Diagnosis: Flat Foot," the log noted solemnly. On 6 December at 1030, a ship's cook second class returned to the ship from Nanakuli Beach for "diagnosis and medical treatment." By 1330 he was on his way back to Camp Andrews to complete his temporary duty: tribute indeed to the ship's efficient medical care. At 0730 on 7 December, a gunner's mate, three electrician's mates, and a quartermaster left the ship for duty with the shore patrol in Honolulu.

Fortune was also dealing her hands to the other ships and to the men in them. From 22 to 28 November we were at sea with Task Force One. By noon of the twenty-eighth the five battleships had returned to their berths along Ford Island, with the exception of the *Pennsylvania*. She entered Dry Dock Number One. The *California* would stay at Berth F-3 longer than any of us could imagine. Just as the *West Virginia* would stay at interrupted quay F-6.

Early on the morning of 5 December, the *Lexington* and other units of Task Force Three—the heavy cruisers *Astoria*, *Chicago*, and *Portland* and

Seven radio gang buddies on one of their last peacetime liberties in November 1941. *Bottom row, from left*: Bill Fisher, the author, and "Red" Goff. *Standing*: Johnny Johnson, Al Raphalowitz, Dave Rosenberg, and striker "Kid" Butler. Six of the seven survived the Japanese attack.

five destroyers—began standing out. Their mission: to deliver marine scout-bomber squadron 231 to Midway.

Between 0910 and 1007 on that day, the *Arizona, Oklahoma,* and *Nevada* stood in and moored along Battleship Row. At 1622 the *Utah* stood in and moored on the other side of Ford Island. Three of these four ships would never stand out again.

> *0342. Mine vessel* Condor, *on routine sweep of harbor entrance, sights submarine periscope, notifies picket destroyer* Ward *by blinker light.* Ward *goes to general quarters, fails to find submarine after hour's search, secures from general quarters.*

On my midwatch, the twenty-four-hour clock on the bulkhead above the supervisor's desk marked the sluggish advance of the minutes. Except for the Fox Sked, which droned on implacably with scarcely a break between messages, it was very quiet in main radio. The training circuit had been secured hours before. My 2,716-kc circuit crackled occasionally with static. There were no emergency messages from CinCPac, no messages on the 500-kc international-calling-and-distress frequency, and none on the 355-kc ship-to-shore circuit.

The mighty Pacific Fleet was sleeping and dreaming until reveille. The coffee got blacker and fouler, and Jack Mazeau ordered the striker to make a fresh pot. It was always that way on the midwatch at Pearl Harbor.

Not everyone in the *California* was sleeping and dreaming. Radio two was manned, in case we needed a transmitter in a hurry. So were the signal bridge, the OOD station on the quarterdeck, and main control in the engineering spaces. But seven of the eight boilers were cold; number one was in use only for auxiliary purposes.

The crews of ready machine guns numbers one and two, flanking the armored conning tower, were on duty. Together, they had 400 rounds of ammunition. It was locked up in the ready boxes.

Not quite on duty were the crews of ready antiaircraft guns numbers one and two, located forward on the starboard and port sides of the boat deck. Those men were sleeping nearby. Fifty rounds of five-inch ammunition were in the ready boxes. They too were locked.

Not manned at all were the other guns, from the fifty-calibers to the fourteen-inch main battery; and the ammunition was far below, in the magazines. Nor were sky control, range finders, directors, central station, the plotting room, or the navigation bridge manned. The CXAM search radar was secured, too, since shore returns from the mountains around Pearl Harbor rendered it useless.

Security in port was not our problem. That was the responsibility of the army and its air corps and the Fourteenth Naval District. The army operated an aircraft warning service consisting of five or six truck-mounted SCR-270 radars linked to an information center at Fort Shafter. The air corps conducted an inshore air patrol. The Fourteenth Naval District was charged with inshore ship patrol and long-range air reconnaissance. It also maintained an inner air patrol that covered the fleet operating areas around the Islands. That, essentially, was the joint-defense plan for "impregnable Pearl Harbor." On paper.

*0600. Heavy cruisers* Tone *and* Chikuma *launch four Zero float planes to reconnoiter Pearl Harbor. They report that fleet, with exception of carriers, is in. Shortly afterward, Japanese carriers launch their first wave of forty torpedo planes, forty-nine high-level bombers, fifty-one dive-bombers, forty-three fighters.*

Down on the port side of the first platform deck, we had been on watch for more than five hours. The air was stale and faintly blue from cigarette smoke. We were yawning and fighting boredom. A little more than an hour to go.

"Striker! Cuppa joe!"

*0633. Navy Catalina patrol plane sights midget submarine off harbor entrance, drops smoke pots.*
*0645.* Ward *attacks sub with gunfire and depth charges.*
*0654.* Ward *sends coded radio message to Com 14: "We have attacked fired upon and dropped depth charges upon submarine operating in defensive sea area."*

Messages sent by World War I–era destroyers on picket duty to the commandant of the Fourteenth Naval District were none of our concern in the *California.* (We didn't even monitor that radio circuit. We had plenty of operators now and could easily have done so. But that was not called for in the defense plan.) The old tin cans assigned to Bloch's command spent most of their time rounding up Japanese sampans that wandered into restricted areas and reporting gray whales and schools of dolphins that produced false contacts on the sonar gear.

As we neared the end of our long midwatch, we were not aware of the drama being enacted just outside the harbor entrance. And even if we had been, it probably would have made no difference. Had we copied the *Ward's* message, the watch supervisor would have sent it by pneumatic tube to the communication watch officer in his "sweat box." Since the *California* was not an action or even information addressee, the message would have been of no concern to him, and he would probably have tossed it into his "In" basket. Or, since the message had an "urgent" designation, he might have decided to decode it. But then, to whom would he have sent the plain-language text?

Admiral Pye was at the Halekulani on Waikiki Beach. Most of his staff were also ashore. Captain Bunkley was at the Halekulani. Absent from the ship as well were Commander Stone, the executive officer, and Lieutenant Commanders Dawson, the gunnery officer, and Naquin, the engineering officer. The senior ship's line officer present was Lieutenant Commander M. N. Little, first lieutenant and damage-control officer. The only staff officer above the rank of lieutenant commander on board was Captain Harold C. Train, Pye's chief of staff and aide. And what would he have done? Doubtless, what Pye or Bunkley would have done, and what Kimmel, in the event, did do: ask for verification.

For Kimmel—and for the rest of us—"verification" arrived all too soon.

*0702.  Two army privates at mobile radar station pick up large flight of planes 137 miles north and approaching Oahu, report it to Fort Shafter's information center, are told to forget it. A flight of B-17s is expected in from West Coast.*

During that long midwatch events moved with the inexorability of Greek tragedy. In the *California*, ninety-eight men were sleeping the last hours they would ever sleep. For many of sixty-one others, these would be the last hours they would sleep without pain.

In radio one we walked around our chairs and swallowed more bitter coffee and looked at the clock. Our reliefs were now at breakfast. Let's see, Sunday was hot cakes day. That, at least, was an improvement over our usual fare of "SOS" or dehydrated eggs or navy beans.

At 0715 our reliefs began clattering down the ladder. I caught up my log, turned over the headset, and carried my cup into the comm office. Now for some chow and some shut-eye!

From what I could see through the main-deck ports, it was a typical Sunday morning at Oahu, warm and sunny, with an eight- or ten-knot north wind ruffling the waters. Without having to look, I knew there were banks of cumulus clouds over the Koolaus.

Despite the many cups of coffee I had drunk, I was hungry. After a full breakfast of cereal, hot cakes, and syrup, skipping only the cold toast, I toyed with a final cup of coffee over a cigarette.

*0730. When breakfast truck arrives, Privates Lockard and Elliott secure mobile radar station. Flight of planes they tracked is about 30 miles away.*

I went to compartment B-511-L, got my toilet kit, and made a trip to the head, one deck above. I had just returned and was about to stow the toilet kit in my locker and turn in when it happened.

CHAPTER 12

# The Desperate Hours: "Battle Stations! This Is No Drill!"

*This fell sergeant, death, is strict in his arrest. . . .*
*Hamlet*

*Bong! Bong! Bong! Bong! Bong!* went the general alarm.

From the quarterdeck came the voice of the bosun's mate of the watch over the ship's PA system: "Now hear this! All hands man your battle stations on the double! This is no drill! Repeat: All hands man your battle stations on the double! This is no drill!"

As much from the edge of stress in the normally impersonal voice as from the words, I knew immediately that a catastrophe of some kind was taking place topside. Beyond exchanging surprised looks with several of the radiomen in the compartment, I didn't hesitate. I tossed the toilet kit into my locker, slammed the door, and started for the maintop.

As I left the compartment, I noticed that one man was still sacked out. Since he was a well-known liberty hound, I was sure he had been ashore Saturday evening and was hung over.

"General quarters!" I shouted. "Hit the deck!"

"Another goddam drill," came the muffled reply. "The hell with it." He turned over and went back to sleep.

I didn't think it was a drill. And it wouldn't have made any difference if I had. As I double-timed down passageways and up ladders, I met many men going in the opposite direction. Some looked alarmed; others bewildered; a few more were protesting bitterly. General quarters at 0800 on Sunday morning was unheard of, even in the *California*.

"Now set Material Condition Zed!" There was more than an edge of excitement in the voice of the bosun's mate this time.

The after repair and first-aid parties were assembling at the crew's reception room as I hurdled over the coaming at the break of the upper deck and came out onto the quarterdeck.

The first thing I saw was a plane painted in aluminum. It was flying low and close, passing down the battle line toward the *Nevada* at the landward end of Ford Island. It was a graceful, single-cockpit fighter; the pilot's

```
                              U.S. NAVAL COMMUNICATION SERVICE
   Whitney—12-12-41—25M              U. S. S. RALPH TALBOT  (DD-390)

  Heading:
             Z F5L 0718302  C8Q TART  O  GR 9 BT


             AIRRAID ON PEARL HARBOR X THIS IS NO DRILL BT
```

Copy of air-raid message from CinCPac as received by the USS *Ralph Talbot* on the 2,716-kc emergency radio circuit. The date-time group is in Greenwich mean time, which translates to o800 Hawaiian time on 7 December. "O" in the heading means that the message carries an urgent priority. Former radiomen and signalmen will undoubtedly notice that "Airraid" should have been copied as two words for the text to tally with the group-count of nine in the message heading. (Courtesy of D. H. MacHaffie, Jr, Pearl Harbor Survivors Association.)

helmeted head was clearly visible through the ribbed canopy. Then I saw the rising-sun symbols on the near wing and the fuselage and remembered our aircraft identification drills.

The plane was a Zero.

*The Japanese were attacking Pearl Harbor!*

Before I could reach the mainmast, a torpedo hit the *California*. A plume of water shot up opposite number three turret, not twenty-five feet away. There was a deep, dull explosion. The ship was gripped, shaken, and shoved sideways in the water, much as if all twelve guns of the main battery had been fired in one full-load salvo.

I lost my balance and tumbled to the spotlessly holystoned teak deck. Breaking the fall with my hands, I was up quickly and ascending the steep ladders to the battle top. Dimly conscious of heavy explosions and the drum roll of machine-gun fire, I plugged in my earphones and got out the message pad and pencil. The circuit was dead, but I was ready. I looked around.

The Pacific Fleet was under attack by more planes than I had ever seen in the air at one time. Curiously, the actual sight was less frightening than my speculations about enemy air raids had been. In my mind's eye, I had seen masses of aircraft coming in wing tip to wing tip. But these sixty or seventy planes were widely dispersed, attacking in small groups from every altitude and quarter.

Torpedo planes in single file roared in at no more than a hundred feet above Southeast Loch, whose waters pointed like a dagger at the battle line. As they reached the channel, they launched their long, slim tin fish. The torpedoes entered the water with a modest splash. Their white wakes were arrow straight as they headed unerringly for the sides of the *Oklahoma*, *West Virginia*, *Arizona*, and *Nevada*.

The planes banked and flashed up and over the ships, perfect targets for our antiaircraft batteries. But only some .50-caliber machine guns and the 1.1-inch "pom-poms" in the *Maryland* were firing as yet. I cursed the fate that had made me a radioman instead of a gunner's mate. If I had a .50-caliber in my hands, I would score some hits! As I watched, one plane was hit. It burst into flame, lost altitude, and plunged into the water near Pearl City.

At the same time, Val dive-bombers were peeling off and striking at the battleships, and at the cruisers and destroyers moored in East Loch. I had seen newsreel films of the German Stukas, which descended in near-vertical dives upon their targets. But the Japanese planes came at us in a steep glide, the wind screeching past their blunt engine cowlings and fixed landing gear in a banshee death wail.

The pilots' aim was very good. Alongside the ships, the near misses sent up columns of water hundreds of feet high, accompanied by muffled explosions and spreading concentric shock rings. The hits were marked by the hideous red flashes of high explosives that ripped through steel as if it were cardboard. I knew that every hit, and many of the near misses, were bringing death to numbers of my countrymen. With the crews widely dispersed at battle stations, that was inevitable.

My headset cord was just long enough to allow me to reach the splinter shield that enclosed the catwalk around the main-battery director. I leaned against the chest-high parapet in numb shock. This couldn't be happening! The entire scene had the flickering, two-dimensional quality of a B-grade war film. Surely the Kates and Vals were models, the battleships mock-ups rigged with harmless powder charges, the men actors feigning pain and fear at make-believe catastrophe. At any moment, the screen would go blank. Then this would again be a typical Sunday morning at Pearl Har-

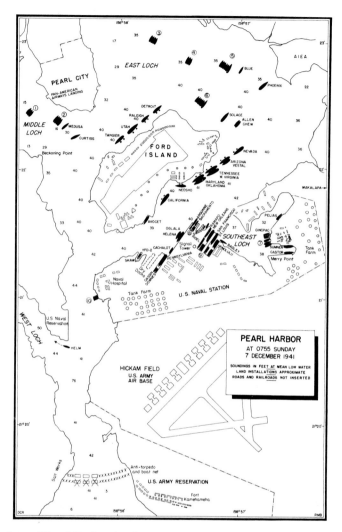

Key to chart of Pearl Harbor (reading NW to SE in nests of ships): 1. Destroyer-minecraft *Ramsay, Gamble, Montgomery*; 2. Destroyer-minecraft *Trever, Breese, Zane, Perry, Wasmuth*; 3. Destroyers *Monaghan, Farragut, Dale, Aylwin*; 4. Destroyers *Henley, Patterson, Ralph Talbot*; 5. Destroyers *Selfridge, Case, Tucker, Reid, Conyngham*; tender *Whitney*; 6. Destroyers *Phelps, Macdonough, Worden, Dewey, Hull*; tender *Dobbin*; 7. Submarines *Narwhal, Dolphin, Tautog*; seaplane tenders *Thornton, Hulbert*; 8. Destroyers *Jarvis* and *Mugford* (inside *Argonne* and *Sacramento*); 9. Destroyer *Cummings*; destroyer-minelayers *Preble, Tracy, Pruitt, Sicard*; destroyer *Schley*; minesweeper *Grebe*; 10. Minesweepers *Bobolink, Vireo, Turkey, Rail, Tern*. Other auxiliaries, not shown, were moored up West Loch. There were also several tugs and yard craft, not shown, in the area of the chart. (Drawn by Robert M. Berish for *The Rising Sun in the Pacific*, courtesy of Naval Historical Center.)

bor—church being rigged under the forecastle canvas for Chaplain William A. Maguire's Catholic service at 0900 and Chaplain Raymond I. Hohenstein's Protestant service at 1000; liberty launches and motor boats shoving off for Merry Point and the officers' club; and men lounging around the forward turrets reading the funny papers and writing letters home in the comfortable assurance that most of the officers were ashore.

But as I stared in disbelief, the nightmare film continued to unreel. A dive-bomber pulled up and leveled off. Its wing guns began to wink

On a typical Sunday morning at Pearl Harbor, 14 September 1941, Chaplain Raymond C. Hohenstein conducts Protestant church services on the forecastle. On the untypical Sunday morning of 7 December, he helped rescue men from a flooding compartment around number two turret barbette before being overcome by smoke and fumes. Later he suffered shrapnel burns on the quarterdeck. Hohenstein became the first living navy chaplain to receive the Purple Heart. He retired from the navy in 1961 with the rank of captain in the chaplain corps. (Photo courtesy of Raymond C. Hohenstein.)

furiously. Angry red tracers streaked by the maintop. That brought me out of my trance. I tumbled to the deck and wrapped myself around the director tower housing.

Why weren't our machine guns firing? When I got to my feet I discovered why: there was no ammunition in the maintop. The gunners could do nothing but shake their fists at the planes and curse in impotent rage.

The only guns in action on the *California* were ready machine guns numbers one and two. Obviously, the ready boxes had been unlocked or broken open. (I later learned they had been smashed open on orders from Lieutenant Commander Bernstein, who had the head-of-department duty.) But something was wrong with number two, the fifty-caliber on the action side. It had to be reloaded by hand after each round was fired. Without rapid fire, it was about as useful as a slingshot. My previous misgivings about our machine gunners had been well founded, it would seem. Their failure to maintain a ready gun for immediate firing was a foul-up of possibly mortal consequence.

A working party was sent from the maintop to the quarterdeck to get ammunition. There, too, the ready boxes were locked. The locks were forced open. The boxes were empty. We all cursed helplessly. Until Ensign Blair or someone got ammunition to us, we were reduced to the status of spectators at this unparalleled disaster.

"My God!" a voice cried. "Look at the *Oklahoma!*"

A few hundred feet astern, at the gasoline dock, was the *Neosho*. Just abaft the oiler was the *Oklahoma*, moored outboard of the *Maryland*. She had an alarming list from several torpedo hits.

Before our horrified eyes, the list slowly increased. Torpedo planes were still attacking her as she heeled inexorably to port. At least two torpedoes hit above her armor belt, and she seemed to shudder uncontrollably. Suddenly, she was at 45 degrees, more list than any battleship in the world was designed to withstand. Her heavy tripod masts carried her over and she capsized, turning until the masts stuck in the mud of the harbor bottom. She settled at 150 degrees, with only her keel and part of her flat bottom showing. In what seemed a sad, defiant gesture, her starboard screw broke the surface of the water.

There was a loud, shrieking death rattle as the water entered her boilers. I could see men scrambling off her hull on the Ford Island side. Many more men were in the water. Already, motor launches and whaleboats were speeding to the scene. Arriving before them were Japanese planes, which strafed the survivors viciously.

"There goes one-third of BatDiv 2," I said unnecessarily. Hundreds of men must be trapped below decks, doomed to the horrid end all sailors dread so much that none dares mention it. I mumbled a brief prayer that death be swift and merciful for them.

Fifteen minutes before, the *Oklahoma* had been a proud battleship of the U.S. Navy. In all her twenty-five years, including service in World War I, she had never fired her ten fourteen-inch guns in anger. Now, before she could fulfill her intended destiny, she had been treacherously struck down. It was an ignominious and revolting way for a man-of-war and her crew to die.

How could it have happened? American battleships didn't capsize. That, I had been told, was next to impossible. But it wasn't impossible; how could I dispute the evidence of my eyes? She must have been wide open on the third deck and below, I decided, and the crew had had no time to set Condition Zed.

Nor did I have any time for further unprofitable speculation. A torpedo plane—probably one of those that had given the *Oklahoma* the coup de grace as she was turning turtle—was approaching from our starboard quarter on a strafing run. Bullets from its 7.7-mm machine gun in the rear cockpit sprayed the quarterdeck and boat deck below us as I dived for cover again, still tethered to my headset. It never occurred to me to take it off. The circuit might be needed for emergency communications—if one of our planes got aloft and if we still had an operative receiver that could feed its transmissions to the foretop and maintop. Beyond that, leaving one's general-quarters station without being properly relieved was an offense of the most serious magnitude—especially under these conditions.

Back at the splinter shield, I could see that the *West Virginia* had been badly, perhaps fatally, wounded by torpedoes and bombs. Her forecastle was already nearly awash; oily black smoke shot through with flickering carmine highlights was pouring from her superstructure, casemates, and boat deck.

Our sister ship, the *Tennessee*, moored inboard of the "Wee-Vee," seemed undamaged. But less than a hundred feet abaft, the *Arizona* was under heavy attack by dive-bombers.

All at once, a mighty thunderclap of sound, deep and terrible, rode over the cacophony of planes and bombs and now-awakening guns. Concussive waves shook the *California* mainmast to its foundations. From the upper deck and superstructure of the *Arizona* a red fireball shot up and spread into a mushroom of death nearly a thousand feet high. Battleship Row was pelted with fiery missiles of burning debris. Following the explosion, clouds of dense black smoke billowed out. The prevailing northeast trade wind caught them and blew them along toward the *California*.

"The *Arizona* is gone!" a lookout said in awed tones. "My God, a bomb must have got the magazines!"

At this point my belief in my sanity was taxed. Such things simply couldn't happen. American battleships didn't blow up. But as the stinking smoke from burning fuel oil began to drift up and over the mainmast, I

The explosion of the *Arizona's* smokeless powder magazines broke the great battleship in two, leaving the mainmast relatively undamaged. She burned until sundown on 9 December. The loss of life was appalling: 1,104 officers and men. (Photo courtesy of Paul Stillwell.)

knew that the unthinkable was happening. Why and how, I did not know—and I would be very lucky to survive long enough to find out. A spasm of chilling fear knotted my stomach muscles. Could this be the end of a short and inglorious life?

It seemed very likely, for the *California* was herself in danger. The word we got by sound-powered phone was that we had taken three torpedo hits, one forward and two aft. (Actually, there were two: one forward at frame 52, and one aft at frame 101. The time of the nearly simultaneous hits was very early in the attack, around 0800 to 0805.) Water and fuel oil from ruptured wing tanks were flooding the third deck through a large number of open manhole covers to the voids of the torpedo defensive system. (In a classic example of unfortunate timing, the ship was scheduled for a material inspection on Monday, 8 December, and many voids had been opened up in preparation for the inspection party.)

Properly "buttoned up," the *California* could have shrugged off two or even three torpedoes with minor listing that would have been quickly corrected by counter-flooding the starboard voids. Instead, she had assumed a port list of fourteen or fifteen degrees.

The list was still increasing. Suddenly, I found myself sliding toward the low side of the birdbath, which brought me up sharply against the splinter shield. My earphones were jerked from my head. Before replacing them, I looked down. A hundred feet below me was nothing but dirty, oil-streaked and flotsam-filled water. Lifeless bodies from the *Oklahoma* floated face down. Motor launches were crisscrossing the channel, picking up swimming sailors.

If the *California* capsized—and that, I could see, was a distinct possibility—I had at least a fighting chance to join the swimmers. My shipmates below decks had none.

I planned to climb to the opposite side of the splinter shield as the ship went over and launch myself in a long flat dive when the maintop touched the water. If I could avoid getting fouled in the yardarm rigging or the radio antennas, I just might get clear.

As I mentally prepared myself for these desperate actions, I looked across the channel to Ten-Ten Dock. The new light cruiser *Helena* had been torpedoed amidships by a plane that had launched its fish from dead ahead of the *California*; she was listing to starboard and emitting clouds of grayish smoke. Tugs were frantically trying to clear the minelayer *Oglala* from the outboard side of the cruiser and move her to a space alongside the dock abaft. The torpedo must have passed directly under the shallow-draft *Oglala*, but she seemed to have been hit, too. She was listing to port and threatening to pin the *Helena* against the dock.

Up ahead, in Dry Dock Number One, the *Pennsylvania* still looked undamaged. Like the inboard battleships along Ford Island, she was protected from the deadly torpedo attacks that had already crippled the battle line and killed hundreds, probably thousands, of sailors.

Behind the navy yard on the ocean side, rising columns of smoke marked the location of Hickam Field. Off our starboard bow, the seaward end of Ford Island was a shambles of burning hangars, blasted carrier planes and floatplanes, and wrecked Catalina patrol craft. So far as I had seen, not a single plane had become airborne. Where the hell, I wondered bitterly, was all the long-range reconnaissance that was supposed to warn us of an impending attack? How would our admirals ever explain their stupidity?

Directly across the island from me, at Berth F-11, a long gray hull was all that showed of the mobile target ship *Utah*. Like the *Oklahoma*, she had quickly succumbed to torpedo wounds. Commissioned as Battleship Thirty-one in 1911, she had never fired a shot in anger either.

"Poor old *Utah*," a shipmate said in a solemn voice, almost as if he were delivering a eulogy. "She was just in the wrong place at the wrong time. That's where the flattops usually tie up!"

Just abaft the remains of the *Utah*, the old light cruiser *Raleigh* (CL 7) had also been torpedoed and was listing dangerously to port. The four-stacker scout cruisers had little reserve buoyancy; I was sure she would shortly turn turtle. But strenuous efforts were under way to save her, for the crew was jettisoning everything on topside that could be pried loose. The *Raleigh* appeared to have one thing that was in short supply so far: leadership. (The *Raleigh* was ably commanded by Captain R. B. Simons. Unlike so many of the ship captains, he was on board when the attack began. His direction of remedial actions provided the slim margin by which his ship was saved.)

"Listen to this, men!" our talker exclaimed. "We're counterflooding. The list is down to twelve degrees!"

Sure enough, our list had decreased perceptibly. I breathed a huge sigh of relief. Thirty minutes may have passed. But I couldn't trust my normally accurate internal clock this morning. It might have been fifteen minutes or fifty.

"High-altitude bombers overhead!" a lookout with binoculars reported.

I squinted up. Heavy smoke from dozens of fires was darkening the sky. Above it, patches of blue showed amid the drifting cumulus. The planes were at 10,000 or 12,000 feet, looking smaller than birds. They were flying over the battle line in a single long column from the seaward side.

At last, several of our five-inch, twenty-five-caliber antiaircraft guns had ammunition. (It had been passed up by hand from the magazines at great sacrifice, I later learned.) They opened fire with an earpiercing *crack!* The sky was dotted with black puffs of exploding shells from the *California* and many other ships, mostly below the Kates.

Bombs began to fall—metallic specks that reflected the sunlight fitfully as they wobbled down. The specks grew larger and more ominous. I felt totally helpless. These might well be the last few seconds of my life. Whether they were or not depended on the skills of an enemy pilot and bombardier, not on anything I could do.

It was small consolation that I would not die alone. If I had to die, it should be with friends like Johnson or Fisher. I had fine shipmates but no close friends in the maintop.

"Take cover!"

Once again I found myself hugging the rough gray steel of the director housing. Above the continuous din of battle, I heard a series of devastating explosions, each one louder than its predecessor, as the sticks of bombs fell nearer and nearer the ship. The last explosion must have been very close, for the *California* trembled in shock.

"They're gone!"

Slowly and reluctantly, I rose from the deck. I wanted to stay where I was, close to that solid, comforting steel barrier. But both duty and curiosity demanded I get up. At least four intersecting shock rings spread out from points of impact off the port bow. Subsiding geysers of dirty white water drenched the forecastle with spray.

"Near-miss damage," our talker told us. "Flooding at the port bow, first platform deck and below."

Other battleships had been hit, but there was no time to assess the extent of their damage.

"Here the bastards come again!"

The bombers had reversed course and were making another run, as calmly as if they were at target practice. Every gun in the harbor that had ammunition opened up again, to no effect that I could see. The Kates were above the range of our five-inch twenty-fives, at least under these chaotic conditions, with no adequate director control. Once more, death yawed down from 10,000 feet in a high eerie whistle. Again, the director housing was my only refuge. Even my prayers seemed lost in the hellish racket of the guns.

There were four explosions this time, not quite as close as the first ones. Back at the parapet, I saw that the bombs had hit the lagoon between the ship and Ford Island, two of them just forward of the mainmast. Below me, the Blue Goose, a sitting duck on its catapult atop number three turret, had been holed with shrapnel.

The bombardiers had now achieved near misses on both sides of us. Either they had overcompensated after their first run or the smoke from the burning *Arizona* and *West Virginia*, which now partially shrouded our ship, had spoiled their aim. At least they're human, I told myself grimly.

At that precise moment, a dive-bomber came plummeting down through the smoke and released from less than a thousand feet. The bomb hit the upper deck on the starboard side, penetrated, and detonated below with a heavy explosion that seemed to lift the *California* bodily from the water. A sheet of flame shot up past the navigation bridge, followed by dense smoke. The bomb had struck just a little forward of the ship's service area, one deck below. I felt a sickening certainty that casualties had been heavy.

Now there was a lull in the action. I was sure it was only temporary. I peered through the smoke, trying to spot Fisher in the foretop. To my great relief, he was there: the earphones were positive identification. I waved; he waved back. I didn't have time to reflect on the fantastic series of events that had brought two high-school buddies from a small town to battle stations in the foretop and maintop of the same battleship on 7 December 1941. ("Battle stations" seemed a gross misnomer on this Sunday. There was nothing to do battle with. "Slaughter stations" would have been more appropriate.)

Flags and pennants were flying from the starboard and port yardarms. Someone from the flag command was on board and had directed Task Forces One and Two to sortie.

What a crock that was, I thought. The *Oklahoma* had capsized, wedging the *Maryland* against interrupted quay F-5. The *West Virginia* was either sinking or already on the bottom, trapping the *Tennessee* against Berth F-6. The *Arizona*, at F-7, had blown up. The *Nevada*, at F-8, had been torpedoed and bombed. The *California* was burning, listing, and sinking. The *Pennsylvania* was sitting on her keel blocks in Dry Dock Number One. Eight of the nine ships of the battle line were out of action; and the only reason it wasn't nine of nine was that the *Colorado* was in Bremerton. All of this had taken place within thirty-five or forty minutes of the first torpedo and dive-bombing attacks.

Turning to look aft, I saw that the venerable repair ship *Vestal* had cleared the blazing *Arizona*, to which she had been moored bow to stern, and was being moved toward Aiea at a snail's pace by a couple of yard tugs. She was down by the stern and flooding. Apparently, she intended to beach in the shoal waters off Aiea Landing. The miracle was that she had survived the monstrous explosion alongside. Thank God her bridge structure had been facing away from the direction of the magazine detonation.

Now there was movement from the *Neosho* at the gasoline dock. Her lines had been cut with axes and she was slowly backing away. She barely cleared the upside-down *Oklahoma* and gained sternway past the battle line. I had a very personal interest in the *Neosho*, the first ship I had gone to sea in. I watched her ring up full ahead and clear the channel unscathed, turning into Southeast Loch to the relative safety of Merry Point. She was low in the water, undoubtedly from a large cargo of aviation gasoline. If the Japanese had hit her at the dock, the resulting holocaust probably would have incinerated everyone above decks on the *California* and *Maryland*. Including Mason, T. C., RM3, USNR.

I looked toward East Loch, where a destroyer had managed to get under way and was proceeding slowly toward the channel. She was one of the *Farragut* class of 1931–32, easily recognizable by the double forecastle break. An observant spotter told us that she was the *Monaghan* (DD 354). As she came into the channel opposite Pearl City, I witnessed another astonishing sight. The conning tower of a small submarine was broaching directly ahead of her, and only a few hundred yards abeam of the aircraft tender *Curtiss* (AV 4).

As we crowded to the parapet to watch, the sub launched a torpedo at the *Curtiss*. Its bubbly wake churned past the tender; the torpedo detonated against a dock at Pearl City. I couldn't imagine how the sub could have missed at that range.

The *Monaghan* attacked while gathering speed. She opened fire with her number one five-inch gun at maximum depression. The range was so short that the shell passed over the sub's conning tower and ricocheted off a derrick barge tied up at Beckoning Point. The barge began to burn.

Coming on swiftly now, and with no maneuvering room, the *Monaghan* rammed the submarine. The destroyer's bow rode up over the enemy's conning tower. As she passed the sub, a couple of depth charges were rolled off her fantail. Almost immediately, two explosions sent up boiling fountains of oily water that seemed to lift the *Monaghan's* stern clear out of the water. The sub sank from sight while we shouted and embraced and slapped each other on the backs. Aside from the very few Japanese planes that had fallen, this was the first thing we had had to cheer about.

The *Monaghan's* daring attack had put her in jeopardy. Unable to halt her forward progress, she fouled the burning derrick barge. Her stem drove right up onto the beach. After an anxious minute or two, she backed off, apparently with only minor damage, and stood down the channel toward the sea.

"Beautiful, beautiful!" "Well done, *Monaghan!*" "Let's get us another one!" we shouted hoarsely after her.

Belatedly, some of our flag staff began returning to the ship. The admiral's barge came alongside the officers' accommodation ladder. I recognized the short broad form of Admiral Pye in civilian clothes leading several subordinates toward the bridge, including Captain A. E. Smith, his operations officer, and Commander B. H. Hanlon, his gunnery officer. Captain Bunkley had not yet put in an appearance. Nor had any of the other missing ship's officers, so far as I had seen.

I thought about our philosopher second class. If he lived through the morning, what would he say now about our "two Jonahs," I wondered numbly.

But, whether or not he was a Jonah, Bunkley should have come back to his ship by now. The Japanese were returning with a second huge wave of horizontal bombers, dive-bombers, and fighters.

In another example of the luckless timing that doomed our forces, a squadron of Dauntless dive-bombers from the *Enterprise*, which was returning from a plane delivery mission to Wake Island, picked this moment to land at Ford Island. Three or four were shot down by the Japanese as they approached Oahu, I discovered later; the other twelve or thirteen ran a gauntlet of antiaircraft fire and managed to land safely on the airstrip.

I saw them coming in low and fast, their frantic recognition signals ignored, the white stars on wings and fuselages clearly identifying them as American, and marveled that all of them made it. By now, most of the ships had ammunition and were firing at any target of opportunity. To our gunners, all planes were enemy.

"The *Nevada* is standing out!"

By some miracle, the *Nevada* had got up steam, cleared her berth abaft the *Arizona*, and was slowly moving past what had been the battle line toward the sea. As she approached the *California*, I could see that she was down by the head from a torpedo hit well forward. She showed bomb damage on her foremast and quarterdeck. Undaunted, she stood down the channel against a stygian backdrop of billowing black smoke.

Seeing her under way, all the dive-bombers over Pearl Harbor switched from their assigned targets and attacked the *Nevada*. They came down in swift, steep dives, releasing their bombs when they were only a few hundred feet above the *Oklahoma*'s sister ship. Giant curtains of water from near misses hid her from view for long moments. A bomb hit forward, near her number one turret. The *Nevada* rose up from the water and fell back heavily, as if she had struck a mine. Another palpable hit somewhere amidships sent debris flying over her masts in a fearful red spray.

As she passed between Ten-Ten Dock and the *California*, her entire forecastle, bridge, and superstructure areas were ablaze with yellow flame. Fresh explosions sent scarlet tendrils shooting up through the yellow mass. All her antiaircraft guns were firing, some in the very midst of fire and smoke. At her stern, the Stars and Stripes flew proudly from the flagstaff.

I had never seen anything so gallant. I choked with emotion as she came bravely on. A Val was hit and plunged into the water at full dive. In the maintop, we gave the crew of the *Nevada* our shouted encouragement and our prayerful curses and our great vows of vengeance. Some of us were crying unashamed tears. In that moment, which would live in history, I understood why men of a special kind were willing to die for abstractions named Duty, Honor, Country.

Ahead of the *Nevada*, a large pipeline snaked out from Ford Island in a semicircle ending at the dredge *Turbine*. Since it blocked more than half the channel, the line was always disconnected and pulled clear when the battleships were scheduled to stand out. This morning, of course, it was still in place. But the sailor conning the *Nevada* squeezed her between the dredge and the dry-dock area without slowing down.

The flames were now shooting up past her antiaircraft directors nearly to her foretop. She had been hit repeatedly, and Pearl Harbor was pouring into her hull; her bow was low in the water. If she were to sink in the channel, she would plug up the entire harbor like a cork in a bottle. With bitter regret, we watched her run her bow into shallow water between the floating dry dock and Hospital Point. The current carried her stern around, and she finished her evolution pointing back up the channel she had tried so valiantly to follow to freedom.

The sortie of the *Nevada* had failed. It remains the most magnificent, heart-stopping failure I have ever seen.

Now the planes turned their attention to the still-undamaged ships on the navy yard side. While the high-level bombers crossed and recrossed over their targets, the dive-bombers concentrated on the dry-dock area. The people who had told me the Japanese couldn't fly combat aircraft because of their defective vision were either victims of wishful thinking or were fools, I thought, while anger and fear mingled with grudging respect. My God, they were superb!

Bombs hit all around the floating dry dock that held the destroyer *Shaw*. All at once the shock waves of another mighty explosion assaulted our eardrums as they shook the mainmast. A mountain of ocher flame built itself against the sky. From the boiling flanks of the expanding mass, jagged ribbons and streamers of red and white pinwheeled out. The *Shaw* had blown up, obviously from a direct hit on a magazine. The floating dry dock was sinking and taking her down with it.

Just when it seemed there was no end to the horror, comic relief of a sort was provided by the *Aylwin* (DD 355). Moored to the same "nest" from which her sister *Monaghan* had departed to sink the midget submarine, the *Aylwin* had finally got steam up and cast off her moorings. As she eased into the channel, she was pursued by a motor launch moving at full throttle. From the bow of the launch, an officer was desperately trying to wave her down.

One of our lookouts put his glasses on this scene.

"She's being conned by a junior officer," he reported. "Prob'ly an ensign. He's looking aft. The officer in the launch—he's gotta be the skipper—is having a fit. He damn near fell overboard! The ensign is turning away. He's giving an order. The goddam tin can is picking up speed! She's leaving the S.O.B. behind!"

Despite the death and destruction all around us, and the bombers above, we laughed. Uproariously.

But the laughter did not last long. From Dry Dock Number One, the destroyers *Cassin* and *Downes* (DD 372 and DD 375) were blazing from forecastle to fantail. Now a series of explosions sent flames and shrapnel flying in a deadly fireworks display. When next I looked, the blackened *Cassin* had rolled over onto the *Downes*. Someone had ordered the dry dock flooded, but the oil rose with the water and was burning around the destroyers and the *Pennsylvania* astern. Sailors who had abandoned ship and yard workmen were fighting the inferno from dockside, even while they ducked the enfilading fire of broken, hot metal.

When I had last checked Ten-Ten Dock, the *Helena* was there, with the *Oglala* astern. Now there was one ship. The *Oglala* had capsized against the dock. Only an arc of two-toned steel that was her midships plates showed above the water. Her starboard accommodation ladder was still intact, now in an oddly horizontal position.*

---

*It was the pressure wave from the torpedo which struck the *Helena* that undid the thin-skinned *Oglala*. The old girl's crew was unable to isolate the flooding that occurred through ruptured shell plating. The joke among navy enlisted men was that the *Oglala* had died of fright.

In this classic combat photo, the destroyer *Shaw* blows up in her dry dock from a direct bomb hit on a magazine. The two forward turrets of the beached battleship *Nevada* can be seen at the far right. (Official U.S. Navy photograph.)

Across Ford Island, the tender *Curtiss* was giving a better account of herself than the leaderless and nearly ammunitionless *California* was. A Val whistled down and unloaded a bomb, probably intended for us, that dug a crater on Ford Island. As the plane pulled up, it was hit by concentrated fire from the *Curtiss*, the cruiser *Phoenix*, and several destroyers. Smoke trailed out aft. A dull orange glow spread and flared up and devoured the dive-bomber. As it began to break up, it hit the starboard seaplane crane of the tender with a *whump*, bounced off, and spun onto the main deck. Burning parts and gasoline sprayed in all directions.

Now the *Curtiss*, like a valiant schoolboy besieged by a gang of bullies, came under attack from several dive-bombers. I couldn't understand why they preferred a seaplane tender to a crippled battleship. The *Curtiss* was straddled with near misses. A mushroom of scarlet proclaimed a hit in the hangar area. Her machine guns kept chattering away. The tracers probed for and bit into one Val. It disintegrated in a miniature fireball. Only a rain of unidentifiable bits and pieces marked where it had been. Once again, we helpless spectators cheered.

"The fires below decks are out of control. They're running toward the powder magazines!"

Our talker passed along this ominous intelligence. His voice was shaky and he looked scared. I didn't blame him. In all our minds was the dreadful fate of the *Arizona*. We didn't see how anyone could have got off alive.

We knew the *California* was burning because brown smoke was puffing up from amidships and drifting past the superstructure. Apparently, the bomb hit had set off a number of fires that were spreading fore and aft.

It did not take much imagination to visualize the scene below: the broken bodies, the blasted machinery, the choking air, and the encroaching water and oil and fire, all within a steely darkness illuminated only by the dim battle lamps.

The torpedo that impacted forward must have been in the vicinity of main radio. I thought about that dogless hatch cover and prayed that someone on the third deck had survived to let Johnson and Reeves and A. Q. Beal and Jack Mazeau and Joe Ross and all the others out. Of course, if the magazines went, it wouldn't make any difference.

But fear for my friends in the radio gang was swallowed up by the terror of a renewed attack.

"Planes coming in on a strafing run astern! Take cover!"

Fighters and dive-bombers roared out of the smoke at just above masthead height. Tracers sprayed the quarterdeck and boat deck and forecastle.

Once again I was on the deck, clinging to the director housing. I knew full well that the thin outside splinter shield offered no real protection against machine gun and cannon fire. One might survive 7.7-mm-gun wounds if they weren't in a vital place. But not wounds from a 20-mm cannon. I made myself as small as I could. It wasn't nearly small enough. Once I had been the hunter. Now I was the hunted. I swore that if I were permitted to live, I would never again kill an animal.

When the attack shifted to ships on the navy yard side, I stood up gingerly but gratefully. To my surprise, no one in the maintop was hit. But below me, on the quarterdeck and boat deck, a number of men were down. Some of them were lying very still, in what I recognized immediately as the loose, crumpled postures of death. I thought I knew two of them, but didn't want to believe it. Others were being assisted to cover and to the after dressing station in the crew's reception room.

I looked aft, to see how the other battleships were faring. I couldn't see any. The fuel oil that had poured from the stricken *Arizona* and *West Virginia* by the hundreds of thousands of gallons had formed a viscous sludge on top of the water. It had caught fire. Orange flames rose into great ebony clouds. The prevailing wind was pushing the fire toward us. Passing inshore from the *Oklahoma*, it had already engulfed the *Tennessee* and *Maryland*. Unless the wind shifted, it would very soon be at our stern.

Now there was noisy movement from below. Panting from their exertions, two seamen emerged from the ladder. Each was carrying a box of loose fifty-caliber ammunition. Meanwhile Ensign Blair had returned to the maintop. (I learned later that he had directed a ten-man working party in bringing up machine-gun ammunition from the forward torpedo hold on the second platform deck. Some of it had finally reached the two ready guns; a number of the bearers were killed when the bomb struck the starboard side as they were heading for the foretop and maintop.)

We were threatened by a magazine explosion, by strafing planes, and by fire on the water. Two seamen had risked their lives to bring us two boxes of loose ammo with which to defend ourselves. There was something terribly ironic about that; something piteous; something very admirable.

Since we had no clipping machine, everyone pitched in to belt the ammunition by hand. Now, at least, we would be able to fight back! Only a hundred or so rounds had been belted and the gunners were at their handlebars, impatient for action, when the word was passed over the sound-power phone.

CHAPTER 13

# The Desperate Hours: "Abandon Ship!"

*Death is lighter than a feather . . . duty is heavier than a mountain.*
Motto of Japanese Second (Sendai) Division

"Abandon ship!"

Our talker's face was the color of defeat. His voice trembled.

I took off my headset and looked aft. The oil fire had reached the *California*. Flames were running along the port side of the quarterdeck. The smoke was so dense I could scarcely see.

The shadow of panic began to undermine the resolve of the men in the maintop. The seamen stopped belting ammunition and started crowding toward the ladder.

"Belay that last word! Back to your stations! Repeat: Belay that last word!"

All hands looked at the talker in fear, wonder, and various gradations of hostility.

"The hell with that!" one man muttered. "Let's get the hell out of here!"

It was a tense moment. But there were leaders present. Reluctantly, with many an apprehensive glance at the skies above and the fire at our stern, the men went back to the ammo belts.

I had wondered if the captain would know what to do in a real emergency. Faced now with his most important command decision, it appeared that he couldn't make up his mind.

Very soon the word came again, "Abandon ship! Repeat: Abandon ship!"

Carefully, I unplugged my headset and stowed it away with the unused message pad and pencil. I felt very calm as I joined the exodus from the maintop. Halfway down, the seaman behind me on the ladder began pushing at me impatiently.

"Christ, let's hurry it up!" he said in a thin, fearful voice. As I turned, I could see that his eyes were wide and unfocused.

"Take it easy, mate," I said soothingly. "We've got plenty of time. No problem." That put an end to his brief loss of control. I learned then that hysteria must be dealt with immediately, and coolly, before it spreads and becomes general.

The crew of the *California* abandon ship at about 1000 hours on 7 December 1941. Oil from the *Arizona* and *West Virginia* is now burning along nearly the full length of the port side. Signal flags belay a previous order for ships of the two task forces to sortie. At the far right can be seen the keel of the capsized *Oklahoma*. (Photo courtesy of the National Archives.)

"Every one put a life jacket on!" was the shouted command.

On the starboard quarterdeck, life preservers had been pulled from their storage space inside the mainmast and piled up in a heap. I slipped into one of the bulky jackets, pulled the tapes snug, and tied them off, as I had long ago been instructed.

My mates of the maintop had scattered; I would never see any of them again. I was alone at the abandon-ship station—and that is to be very much alone. *Abandon ship*—those two words turn a disciplined crew into a mass of individualists, held together by the tenuous bonds of common loyalty, tradition, and courage. Without these, the mass becomes a mob. All around me, men were donning life jackets in a hurried but orderly way and heading for the edge of the quarterdeck.

Most of the crew were abandoning over the forecastle, jumping into the water or sliding down the mooring lines to the forward quay. Some were using the rope ladders suspended from the boat booms. Just to my right, the wounded were being carried down the officer's accommodation ladder to waiting boats.

As I stood by the lifeline, I could see that the oily water was beginning to burn around the starboard fantail. Through the smoke appeared the dim outline of a small vessel, a yard tug. Passing close astern, the tug backed down toward Ford Island. This action by an unsung hero put open water between the swimming men and the flames.

While others splashed into the lagoon and started for the island, I hung back. It was a couple of hundred yards at best. I wasn't concerned about that. With *California*'s list, the water looked a long way down. I wasn't concerned about that, either.

A strange passivity had suddenly gripped me. I was reluctant to abandon my ship. My home. It seemed such a cowardly thing to do. I felt strangely unimportant and ineffectual. My ship was going down in a furnace of flame along the port side, she might blow up at any moment, and I had done nothing to save her. Rather than perform deeds of valor, I had worn a pair of useless earphones plugged into a dead radio circuit for two hours of inert horror. I felt I would wear a brand of failure and humiliation all the rest of my life, like a scarlet letter.

Unless I could redeem myself. And, thereby, help my navy redeem itself from its shame. I tore off the life jacket, threw it to the deck, and removed my shoes and socks. They would only slow me down. Making sure there were no bobbing heads below, I jumped feet first into the oily mess.

When I came to the surface, gummy black oil was clogging my nose and ears, burning under my eyelids. The rank, sweet taste of the stuff made me want to vomit. Striking out for shore, I passed a number of slower swimmers, but no one needed help. Soon I was standing on Ford Island, wondering what to do next. Again, I seemed to be quite alone.

I couldn't see any planes. Was it possible that the air raid was over? I was not going to be surprised a second time; I began looking for a place of refuge. Nothing was nearby except a bomb crater, probably a souvenir of the dive-bomber that had badly missed the *California* and been destroyed by the guns of the *Curtiss* and the *Phoenix*.

"Take cover!"

Faintly, I heard the shouted warning. I ran to the bomb crater and dived in headfirst. My reaction was instinctive: surely lightning wouldn't strike the same spot twice?

The second attack wave had already retired, I later learned; so the Zero that passed low over my crater, its bullets kicking up puffs of dust nearby, must have been a straggler. I huddled in the bottom, consumed with rage. If I had had a gun, surely I would have charged out shooting. The ultimate humiliation would be to survive the destruction of my ship, only to be exterminated like a varmint in its lair on this miserable burning island!

When I crawled out and stood up, I examined myself for wounds. I was black from top to bottom. Some brownish smears against the black had come from the fresh new earth of the bomb crater. A thin stream of blood ran down the shin of my left leg. I didn't know when that had happened; but it didn't seem serious. I went looking for my shipmates.

I found many of them in an open area near the ferry landing. Some were as filthy as I was. A ship's officer I didn't know but vaguely recognized was trying to get them assembled in a military formation. He had on a clean white uniform—one of the late arrivals, no doubt. No one was anxious to offer himself as a clay pigeon for another strafing plane, but the formation was finally drawn up.

We now learned that Captain Bunkley had cancelled his abandon ship and ordered the crew to return. The oil fire had passed down the full length of the port side, but the wind had shifted, blowing the flames clear.

"She's a great ship," the officer encouraged. "If we all go back and fight the fires, we have a good chance of saving her."

We shuffled our feet and looked at one another. If the *California* was worth saving, then why the hell hadn't we stayed aboard and saved her?

But of course I would return and fight the fires. The *California* was my ship; it was my duty. Redemption began here.

We were taken back by motor launch and whaleboat and organized into a long bucket brigade. Stationed on the boat deck, I concentrated on not spilling a drop of the precious water as I swung each bucket along. Just ahead, smoke was boiling up from the open hatches, the bomb hole in the upper deck, and the casemate gun ports. The boat deck was hot against my bare feet from the fires below. I thought about the powder magazines. No one had told me whether or not the efforts to flood them had been successful. For all I knew, the *California* might blow up at any moment. It

Listing to port and still burning, the *California* settles into the muddy bottom of Pearl Harbor in the aftermath of the Japanese attack. The *California* was well designed. She would have shrugged off the two torpedo hits she received had not many open manhole covers to the voids of the torpedo defensive system destroyed her watertight integrity. (Photo courtesy of the National Archives.)

was a lot easier to be brave, I decided, when you knew what the hell was going on.

A yard tug came alongside and began directing high-velocity streams of water at the fires. The bucket brigade was secured. We had tried hard but accomplished little.

For the second time that day, I abandoned ship.

Soon I was in a motor launch with a number of ComBatFor flag personnel, heading for the submarine base. Being a radioman suddenly made me special: the rest of the *California* crew had been left behind on Ford Island. Other launches, whaleboats, and motor boats were transporting the wounded to Ten-Ten Dock and Merry Point. A few boats were picking up bodies from the water. The corpses were stacked like cordwood in the footlings under canvas shrouds. I looked away. I had already seen enough death.

The sub base was an oasis of calm on the fringes of the carnage. Both it and the tank farm on Makalapa Hill had escaped damage. A reception party was awaiting us.

A pharmacist's mate dabbed at my wound and dismissed it. "That's nothin'!"

Another petty officer directed me to the CinCPac radiomen's quarters in the administration building. "Get cleaned up, buddy," he said grimly. "We're gonna need you."

Solicitous radiomen took me to the head. I stripped off my filthy uniform. I still had my billfold and gave it to someone for safekeeping. With

Smoke from the funeral pyre of the battle line rises thousands of feet high as it drifts toward Honolulu in this photo, taken shortly after the Japanese striking force withdrew. At left is the sinking flagship *California*; in the middle background is the *Maryland*, with the capsized *Oklahoma* alongside. Oily black smoke conceals the sunken *West Virginia* and the broken *Arizona*. (Photo courtesy of Paul Stillwell.)

a towel dipped into a bucket of kerosene, I swabbed off the oil. Then I stepped into a glorious hot shower.

I stayed there a long time. A feeling of elation possessed me. I was alive! The other men in the shower shouted and laughed and sang. I joined them. Suddenly, I felt ashamed of myself. How many of my shipmates were dead, wounded, hideously burned? We had suffered a shocking defeat at the hands of the Japanese Navy. Why was I singing? It took years, and additional combat experience, before I forgave myself for callously celebrating personal survival after the fell sergeant had arrested so many of my comrades in arms.

I was given new skivvies, dungarees, shoes, and a white hat, and was taken to the mess hall. To my great joy, a number of the other *California* radiomen were there: Fisher, Stammerjohn, Red Goff, Jerry Litwak, Al Raphalowitz, A. L. Smith, Joe Goveia, and Jack Mazeau among them. But not Johnson. And not Reeves. I couldn't eat, but I drank pints of coffee while we exchanged experiences.

Jack Mazeau, like me, had been about to turn in after our midwatch; he raced to main radio in his skivvies. Events there, he told me, had been the realization of a claustrophobic's worst fears. The first torpedo struck directly abreast the radio compartment. The inboard torpedo bulkhead held, but the ship immediately began to list. The hatch cover was slammed down and secured. Water and oil soon started to pour into the radio shack from the ventilation trunks.

The tactical radio circuits we customarily manned at sea—such as Admiral Pye's 386-kc command circuit and other frequencies for the task group commanders—were set up by Mazeau. Radio two was ordered to tune transmitters.

Mazeau had been closely questioned by Chief Reeves. Was there any increase in radio traffic on the mid? Had there been any signs of an impending attack? Mazeau told him there had been none. I nodded. I knew there hadn't been any.

An urgent message, "PREPARE TO GET UNDER WAY," was received over 355 kc and 2,716 kc from CinCPac. Main radio retransmitted the message to Pye's Task Force One over 386 kc.

"The *West Virginia* fired right back with a snappy R," Mazeau reported. "My God! Both of us were already sitting on the bottom of Pearl Harbor!"

"We were ordered to abandon main radio about 0830," Mazeau continued. "Of course, we couldn't open the hatch from below. By beating on it, we got some of the repair crew on the third deck to open it. Chief Reeves was the last man out. I was next to last.

"Even when we were busy carrying some of the men up from the third deck after they were overcome by fumes, the chief kept questioning me about what went on during the night."

"What happened to him?" I asked.

"The last I saw of him, he was headed back below," Mazeau said.

The transmitter room was able to stay in operation until the first abandon ship. After main radio was secured, radio two transmitted messages sent down by telephone from the flag conning tower. Not that it made any difference, Mazeau and Goff and Litwak and the others agreed.

"None of us were going anywhere, fer crissakes!"

Red Goff provided the narration with human interest in his sardonic way. When abandon ship was ordered, he led a group of radiomen through the sacrosanct wardroom. Espying the customary bowl piled high with fruit, he stopped and selected a large red apple.

"Help yourselves, men," he said. "All this stuff came from the crew's mess, anyway!"* When he emerged onto the quarterdeck, now aflame along the port side, he was nonchalantly biting into his apple.

But Red Goff's irreverent humor was something to be savored later. Right now, I was anxious to learn about casualties. And my worst fears were soon realized.

Melvin Grant Johnson was dead.

He had joined a large number of crewmen assembled near ship's service, where they were awaiting reassignment after abandoning flooded compartments on the third deck and below. The bomb that hit the starboard side was rejected by the armored second deck. Rebounding, it exploded under ship's service, ripping up the main deck and devastating the area. An estimated fifty men were killed, and many others were wounded.

Somberly, one of the radiomen told me: "I saw him shortly before the bomb hit. He said he knew this was it; he wasn't going to make it. He said good-by, and we shook hands." The radioman shrugged in a gesture sad and fatalistic. "Five minutes later. . ."

My swashbuckling and great-hearted friend was gone. He had been right that morning on the quarterdeck as we left Long Beach, when he said that he would never see the States again.

I could still hear his voice from that Hotel Street bar on our liberty of 5 December as he recited the lines from Ernest Dowson's best-known poem, ending in the refrain, "I have been faithful to thee, Cynara, in my fashion."

And so he had, to his Cynara and to his country. I knew I had and would have other friends. But never a friend like M. G. Johnson.

Albert Quentin Beal was dead.

During one of the last strafing attacks, he had been cut down on the boat deck.

So he was one of those two loose, crumpled shapes I had refused to recognize.

Now he would pursue no more women; drink no more Vat-69 scotch; read no more Nietzsche and Schopenhauer; write no more romantic poetry.

It was hard to believe that A. Q. was dead. He had brains, panache, an unquenchable exuberance. Of all the members of the radio gang, he seemed the most likely to accomplish great things in the future.

But he was, above all, a man of valor. He was on the boat deck, I was told, because that was where the guns were. He died charging recklessly across open spaces. Even in the face of death, he was heedless of consequences. Perhaps if we had had men of his qualities running the ship that

---

*An officers' mess was authorized to buy provisions from a ship's general mess at cost. But no one could have convinced my truculent friend that the supplies were purchased.

morning, we would have taken the *California* out of Pearl Harbor—or taken her at least as far as the gallant *Nevada* got—and erased the shame of losing her at quayside.

Joe Boyce Ross was dead.

He, too, had been felled by topside strafing.

Joe and I were not friends. Although we had shared the same living compartment for fourteen months, we had little to say to each other. I had thought him a vain, foolish fellow, and he quite obviously dismissed me as an overrated reserve who lacked both operating skills and shoreside savoir faire.

When the permanent landing force had assembled that January morning in San Pedro, he had only been half kidding when he wisecracked that the landing force bill included all the expendables—and that nothing was more expendable than a V-3. There was a larger force, which some called God and others blind chance, that determined who was expendable.

I wondered if, in his twenty-one years, Joe Ross had established one close personal relationship. So far as I knew, he had no real friends in the radio gang, only liberty companions. Would he be missed by certain young ladies in Los Angeles and San Francisco and Seattle—or had they already forgotten him?

But death forces one to consider a man's virtues. As a radioman, he had been dedicated, competent, and ambitious. I thought it most probable that he had been topside with A. Q. Beal in the service of the guns. If so he had, in those last minutes, put the welfare of his ship ahead of his safety. Whatever I thought of Joe personally, I had no doubt that he shared one virtue with A. Q. He, too, had been a man of valor.

Alfred A. Rosenthal was dead.

No one was quite sure what had happened to him. Some said he was among the ship's service group that was wiped out by the bomb. Others thought he never got above the flooding and fume-ridden third deck.

I knew his death would be a terrible blow to his family. Many enlisted men came from homes that were broken or might as well have been. Civilian life had been a bleak, unhappy chapter in their lives that they were only too pleased to close. The navy had become their family. But Al Rosenthal, I remembered, had been a youngest and a favorite son. He had seen no reason why life should not continue to favor him as it had always done, an attitude that occasionally put him athwart navy regulations.

But not even a few days in the brig could subdue his high spirits. Ashore, the handsome New Yorker had what not even Joe Ross could boast of in the Honolulu of 1941: a steady girl friend, a dark-complexioned lass of Portuguese extraction.

Without being close friends, Al Rosenthal and I had always got along well. His loss saddened me even more when I reflected that the four big operators and lovers of the radio gang—Johnson, Beal, Ross, and Rosen-thal—were all dead. Did God have something against lovers? Or was it only that their successes in that area made them unduly reckless or improvident in others?

By now the sun had set on the most evil day in the history of the U.S. Navy. No one knew what the casualties had been; no one knew what fresh disasters might be unleashed upon us in the hours ahead. For a Japanese invasion fleet might even now be preparing to land in force on Oahu. Judging by what I had seen from the top of the mainmast, I had no doubt they would be able to do so.

We were informed that Lieutenant Sugg was setting up an emergency radio station on Ford Island and had asked for certain of the *California* radiomen. I was among them. By virtue of having escaped the ship relatively unscathed, I had acquired an important new status. The world, I began to learn that day, is not run by the best people but by the survivors.

We were outfitted with helmets, gas masks, and Springfield service rifles with ammunition belts. Now, at least, I had a weapon in my hands. Before transportation could be arranged, the air-raid siren began its loud twentieth-century dirge.

"Condition Red! Enemy air raid!"

Jerry Litwak and I and another *California* radioman headed for the top of the ad building. By God, this time we were going to shoot back!

Nothing could be seen in the blacked-out harbor but the hellish glow from the still-burning *Arizona* and *West Virginia*. Not even a star twinkled through the cloud cover. Suddenly, a single gun opened fire from the navy yard. A second joined in, and a third. Within seconds, Pearl Harbor had erupted in a chain reaction of antiaircraft fire of all calibers. The red lines of tracers formed transient geometric patterns across the sky.

"Where are they, God damn it?"

"Over there—by Ford Island!"

We saw dim red and green lights approaching the airstrip from the seaward end. Most of the tracers were flashing in that direction. We took aim at the lights and opened fire with the fierce kind of joy that had possessed those brawling sailors at The Breakers in Seattle. The fact that we didn't have a prayer of hitting an enemy plane at that range was immaterial. That civilians on the other side of the island might be endangered never crossed our minds. We were striking back at the foe who had so humiliated us.

One plane exploded. Another fell in flames. The lights vanished. We kept firing.

"Cease fire! They're friendly planes! Cease fire!"

The gunfire abated and then died out as word reached the various navy, marine, and army gunners.

Litwak and I didn't believe the planes were friendly. Who would be so stupid as to send our planes into the teeth of all the guns ringing Pearl Harbor on the night of 7 December? They simply had to be Japanese, we told ourselves, feeling good about this chance to be actively rather than passively defensive.*

During the "air raid" the third radioman disappeared. I had been too busy to pay any attention.

"Did you see that goddam ——?" Litwak demanded as we made our way back to the radiomen's quarters. "As soon as the shrapnel started flying, he hauled ass. Headed for the basement, the yellow bastard!"

"It's been a rough day, Jerry," I offered.

But my friend was not that easily placated. He formulated plans for us to write a satiric poem that would delineate our shipmate's cowardice for all the world to read.

Any joint literary efforts, however, would have to be postponed. We had a more immediate problem: how to thread our way by motor launch through a darkened harbor lined with nervous gunners and arrive in one piece on Ford Island. Thanks to a cautious coxswain, who spent half his time with the engine idling while we listened for boats and other navigational hazards, we made it to that dismal island without so much as a challenge.

Opposite interrupted quay F-3, we stumbled across the *California's* emergency-radio facility. A tall figure approached, peering at us through the murk.

"Ah, Ted. Litwak. I'm glad to see you both," said Lieutenant Sugg. I was not surprised that he was in a starched white uniform. I would have been surprised if he had not been.

"Ted, I know you've been through a lot," he continued. "But I'm afraid we've lost twenty or twenty-five of our radio gang. Are you up to taking over the Fox Sked for a couple of hours while I get some more watch standers lined up?"

After being in the maintop and abandoning ship twice, NPM Fox was nothing. "Yes sir," I replied.

I slid down into a hastily excavated dugout. It had been covered with planks topped by piles of sandbags. A steamer trunk served as a table for the typewriter and battery-operated receiver. A couple of flashlights had been rigged for illumination. Perched on a stool, I took over the watch.

---

*The planes were six Dauntless dive-bombers from the *Enterprise*, returning from a bootless search for Japanese ships. Tragically, four SBDs were shot down; a fifth was damaged. Four pilots were killed.

It was hot and airless in the dugout. Dirt and sand filtered through the chinks in the roof onto my typewriter and me. The Fox Sked ran on as impersonally as ever. The only sign of a crisis was an increase in messages of urgent priority. None was addressed to the *California*. Within a few hours, she had been reduced from Battle Force flagship to a useless, sinking hulk.

From the Chief of Naval Operations came a plain-language message addressed to all navy commands, ships, and bases. I was reminded in dramatic fashion that I would not be a supernumerary for long as I copied it: "EXECUTE WPL 46 AGAINST JAPAN." A few minutes later another message came: "EXECUTE WPL 46 AGAINST GERMANY AND ITALY."

Execute War Plan 46 against Japan with what? I asked myself. Three flattops, some cruisers, and the eight-ball battleship *Colorado*: that is all we had left in the Pacific to fight with.

Meanwhile, the scuttlebutt was as thick as the sand fleas on Ford Island. Sober-faced shipmates brought me each alarming report: arrows had been cut in the cane fields by saboteurs, pointing the way to Pearl Harbor; Honolulu's water supply had been poisoned by fifth columnists; Japanese transports, up to forty in number, had been sighted off Barbers Point; paratroopers in blue coveralls with rising-sun shoulder patches were dropping on Nanakuli Beach and the Nuuanu Valley.

If the last two rumors were true, Ford Island would be the place where the radio gang made its last stand. How apt that its original name was Mokuumeume (the "isle of strife")! Even now, the radio dugout was being ringed with rifle parapets made of sandbags.

I hastily checked my Springfield. The promise I had made in the maintop applied only to animals—not to men. I hoped I would die well and bravely, atop a pile of enemy bodies, as befitted a descendant of the Chevalier de Bayard.

A radioman ducked around the blanket that served as a blackout curtain. "I'm relieving you," he said unhappily. "Lieutenant Sugg wants to see you."

"I have one more assignment for you, Ted," my radio officer said. "I hate to ask, but I need someone I can depend on." He explained that it was necessary (for reasons now obscure) to set up a voice radio watch on the *California*.

I quickly responded, "Aye aye, sir!" Better to fall on the decks of my ship than in the dust of Mokuumeume.

Encumbered with helmet, gas mask, rifle, and two-way portable radio, I was taken alongside by whaleboat. With the ship listing ten degrees and the quarterdeck awash, the port forecastle was low in the water. It was no great task to clamber to the top of number two turret.

But once on the slanting turret, I and my equipment kept sliding toward

This somber photo of the *California* was apparently taken some hours before the author's solitary vigil atop number two turret. (Official U.S. Navy photograph.)

the sinister black waters to port. Around me, the harbor periodically exploded into sound and fury as nervous antiaircraft gunners let go at every aircraft alarm. I propped my helmet atop the earphones for a modicum of protection from flying shrapnel.

The great ship was now dark and lifeless.* She creaked and groaned as she strained against her mooring lines and hawsers, shifting and settling deeper into the mud of the harbor bottom. Below me nearly a hundred shipmates lay in their iron tomb. And the thoughts of a young petty officer were dark indeed.

This was the turret crown where Johnson and I had had so many long evening conversations, where we had slept side by side through the early mornings. Now what was left of him was below, on the main deck. Chief Reeves undoubtedly lay down there, too, on the third deck. And Tommy Gilbert was probably on the first platform deck. Along with many others.

I did not know what time it was, nor did I know what the dawn would bring. Since reveille on Saturday morning, 6 December, I had had two

---

*The fires in her main deck and starboard casemates had been brought under control around 2200. The ship's log for the midwatch of 8 December reports that the *Widgeon* (ASR 1) and two oil barges were moored along the port side and the *Vireo* (AM 52) and yard tug *Nakomis* along the starboard side, pumping flooded compartments. I do not recall any vessels alongside to port. With the *California* in danger of capsizing, that seems most unlikely.

hours of sleep. My eyes burned; my skin crawled from the kerosene bath at the sub base; I ached all over, like a malaria victim. My inevitable reaction to the face of battle had been postponed for just about as long as possible.

After what seemed an interminable time, my headset crackled with the cheerful tones of Lieutenant Sugg.

"Ted, I'm going to send a boat over to take you off. We have reports the *California* is about to capsize, and I don't want to lose that valuable radio gear."

My gentlemanly radio officer had had a rough day, too. I was sure he hadn't intended his words to come out quite that way. Later, they would make for a good anecdote. If I lived long enough to tell it.

How many sailors have ever been the last man off a navy battleship? For that matter, how many have ever abandoned ship three times in one day?

# CHAPTER 14

# Postscript to Pearl

"Ted. Hey, Ted. Reveille, buddy!"

I struggled up through fathomless depths of sleep. Not another mid-watch? When I broke the surface, I was not in my familiar radiomen's compartment. Where was I?

In a trice, I knew. I was lying across three life jackets spread out on the ground between a rifle emplacement and our emergency radio dugout. My Springfield, helmet, and gas mask were close at hand. Now I knew I hadn't dreamed the events of Sunday, 7 December. I didn't have to look across the water to the sunken *California* to remember what had happened.

"Ted, let's see if we can scrounge some chow. Then we'll relieve the watch."

It was Al Raphalowitz, a fellow V-3 who had been transferred from the *Mississippi* seven months before. Intelligent and thoughtful, Al soon proved to be a loyal, sympathetic friend. Near us, members of the radio gang sprawled in sleep or manned the parapets.

Raphalowitz and I joined a long line in front of the mess hall. In our reasonably clean dungarees, we looked ready for captain's inspection compared to most of the men. Some were still in the oil-soaked T-shirts and shorts with which they had abandoned ship; some were in dirty undress whites; others in dungarees; a few in undress and even dress blues; and several in all combinations of these.

Inside the mess hall, the fare was spartan: prunes, coffee cake, and coffee. We were told that rations were limited: only one or two meals a day would be served. Maybe. We drank refills of hot coffee and were grateful to have it. Men talked in low voices or stared into space in silence, still overwhelmed by the events of yesterday. We all knew there was more to come. Martial law had been declared in the territory of Hawaii: the army had already taken over from the figurehead governor, Joseph E. Poindexter. The headlines in the Honolulu *Advertiser*, which somehow had been delivered, screamed that the Japanese had bombed Guam and Wake islands and invaded Malaya and Thailand.

This photo of the author's childhood sweetheart, Bertha Morton, emerged unscathed from the oily waters of Pearl Harbor on 7 December 1941. Regarding it as a talisman, the author carried it with him through the Solomon Islands and Central Pacific war theaters.

I got out my billfold, which one of the radiomen at the sub base had wiped clean. I still had my seven dollars in bills, now dark with oil stains. On impulse, I pulled out the photo of Bertha Morton. Miraculously, it was untouched by the corrupted waters of Pearl Harbor.

"I'll be God damned," I said. "Look at this, Raff. It went through the lagoon with me yesterday."

Raphalowitz examined the picture front and back. "Well, I'll be God damned!" he repeated with emotion.

I put the photo back almost reverently. It was the reassurance I had needed that all was not lost. I regarded it as a talisman and knew I would keep it with me always. (And so I did: through the landing on Rendova, when the *McCawley,* torpedoed while I was alongside in the *Pawnee,* went down in thirty seconds; through Munda, Vella Lavella, and Bougainville; through the invasion of Pelileu; and through the rescue of the *Houston* off Formosa.)

Talismans were badly needed that first week on Ford Island. The line that supplied fresh water to the island had been severed. There was no water for shaving or showering. There was very little food. We slept in the dirt, either alongside our emplacements or in several dugouts near the radio pit. We had lost all our clothing, toilet articles, and personal gear and didn't know when we would get replacements. We looked more like young Bowery bums than members of the spit-and-polish battleship navy.*

In the confusion of the first and second abandon ships, the radio gang had been broken up and scattered, never to be fully reunited. Some two dozen of us were on dugout duty at Ford Island. Many others were at the submarine base or in the *Pelias* (AS 17), a new sub tender tied up there. A few had found their way to CinCPac's just-completed Makalapa Transmitter Station in the hills east of the sub base. And sixteen were reported dead or missing, including Reeves, Johnson, Gilbert, Beal, Ross, and Rosenthal.

Admiral Pye and his staff were with Kimmel at CinCPac.

The harbor was a scene from Dante's *Inferno*, evoking all the horror of the ninth circle. The battle line was frozen in grotesque positions. The *California* was reeling to port; the *Oklahoma* had capsized (working parties were desperately burning and drilling through her double bottoms to release men trapped in upside-down lower decks); the *West Virginia*, still smoking, rested flat on the bottom, with her forecastle awash; the *Maryland* and *Tennessee* were squeezed against their quays and unable to move an inch.

The *Arizona* was the saddest sight of all. The explosion of her eight smokeless powder magazines had opened up the forward part of the ship for nearly 300 feet in a deadly flower of destruction. Most of the interior structure and outer shell had been demolished. The entire bridge and tripod foremast had toppled and now leaned forward at a twisted forty-five-degree angle. From the sunken forecastle, only the three guns of number two turret still poked their muzzles defiantly above water. The ship had burned until sundown on 9 December. Incongruously, the mainmast still stood straight, tall, and unblackened, overlooking the wreckage.

Oil was everywhere: spreading out from the sunken ships in pernicious rings, mottling the water, blackening the shore. Drifting on the slow currents were life rafts, life jackets, pieces of boats, and other flotsam. The air smelled of bunker oil, fire, smoke—and death.

Across the seaward end of Ford Island, the *Nevada* could be seen beached at Waipio Point. Tugs had moved her there shortly after the attack ended to clear the narrow channel. By now, I had heard about the *Nevada's* band and marine guard at morning colors on 7 December. While machine-

---

*Lest any reader think I have indulged myself in an old war veteran's penchant for exaggeration, my memories of conditions on Ford Island have been largely corroborated by David Rosenberg. On the long midwatches in the radio dugout, and later in the flag conning tower, Dave kept a diary (as I should have done). He took the time to provide me with a verbatim tape recording of his memoir.

gun bullets sprayed around them, the band played the "Star Spangled Banner" to the final note and the marines stood rigidly at "present arms." Then all broke for cover. The battleship made her magnificent sortie with colors flying. Once again, I had tears in my eyes as I silently saluted the *Nevada* and her crew.

Very soon the story of the *Nevada's* chief quartermaster added to her legend. With the captain and executive officer ashore, her senior officer was Lieutenant Commander Francis J. Thomas, the first lieutenant and damage-control officer. Leaving his yeoman in charge of central station, he took over the conning tower as commanding officer.

So far, the situation was an almost exact reprise of the *California's*. Our first lieutenant and damage-control officer, Lieutenant Commander M. N. Little, found himself in command for the first fifty minutes of the attack. Before reporting to the bridge, he ordered his chief yeoman, R. M. Baldwin, to take charge of damage-control personnel in central station.

But there all resemblances between the stories end. Thomas had two boilers on the line, instead of the one that normally provided power in port, and the *Nevada* was in a better condition of readiness than the *California*. Despite damage from two or three bombs and flooding from a torpedo hit that tore a huge hole in the bow, he coolly gave the order to get under way. Under strafing fire, a line-handling party led by Chief Boatswain Edwin J. Hill cast off the lines from the quay and swam back to the ship. (Later that morning Hill was blown overboard and killed by the explosion of several bombs while directing the anchoring detail. He won the Medal of Honor.)

At the helm was Chief Quartermaster Robert Sedberry. It was he who maneuvered the ship away from Berth F-8 without the aid of tugs; cleared the blazing hulk of the *Arizona* close aboard; took the *Nevada* down channel beyond Battleship Row; squeezed her past the dredge *Turbine*; and then, under orders from the signal tower, beached her between the dry-dock area and Hospital Point. It was a superb feat of seamanship, performed with élan under the most desperate circumstances. Commander Thomas was to get most of the official credit, but for the enlisted men the real hero was Sedberry.

We of the *California* had our own heroes: three enlisted men and a reserve ensign, all of whom were to receive the Medal of Honor.

Chief Reeves, one of the four, acted precisely as I would have expected. After main radio was abandoned and he had helped some men to safety, he returned to the burning, rapidly flooding third deck. Realizing the desperate need for ammunition at the antiaircraft batteries, he plunged into the smoke and flames with every able-bodied man he could muster. It was there on the starboard side of the third deck, less than fifty feet from the entrance to the radio shack he had supervised so long and so ably, that he collapsed and died. His citation reads:

For distinguished conduct in the line of his profession, extraordinary courage and disregard of his own safety during the attack on the Fleet in Pearl Harbor, by Japanese forces on 7 December 1941. After the mechanized ammunition hoists were put out of action in the U.S.S. *California*, Reeves, on his own initiative, in a burning passageway, assisted in the maintenance of an ammunition supply by hand to the antiaircraft guns until he was overcome by smoke and fire, which resulted in his death.*

Warrant Gunner Jackson C. Pharris, in charge of the ordnance repair party, was another hero of the third deck, and the only Medal of Honor winner who survived the attack on the *California*. When the first torpedo hit abreast main radio, he was hurled against the overhead by the concussion and fell wounded. As soon as he recovered consciousness, he (assisted by Chief Reeves) set up the hand ammunition train from the five-inch antiaircraft magazines on the second platform deck to the boat deck. As the ammo passers below were overcome by fumes, he recruited replacements. Despite his wounds, he repeatedly dragged unconscious men from compartments that were rapidly filling with oil. His citation concludes, "By his inspiring leadership, his valiant efforts and his extreme loyalty to his ship and her crew, he saved many of his shipmates from death and was largely responsible for keeping the *California* in action during the attack."†

As much as we admired the gallantry of Chief Reeves and Gunner Pharris, the ultimate act of courage and self-abnegation was performed by Machinist's Mate First Class Robert R. Scott. His battle station was the air-compressor compartment, just forward of crew's space A-518-L and number two turret barbette, in the death zone of the third deck. Scott's compressors, vital for the continued service of the guns, supplied the high-pressure jets of air that cleared the five-inch antiaircraft rifles of burning debris after each round was fired.

With the first torpedo hit, oil and water poured up through the open voids. Scott's compartment soon began to flood. The other machinist's mates were forced to evacuate. But Scott refused to leave.

"This is my station," he shouted after them. "I'll stay and give them air as long as the guns are going!"‡

Reluctantly, his division mates slammed the watertight door on the compressors, on the rising oil and water—and on one of the bravest men any sailor ever served with.

The only officer to join this illustrious company was Ensign Herbert C. Jones, USNR, junior officer of the Sixth (Starboard) Division. Joining the Pharris-Reeves ammunition gang, he was supervising the passing of five-

---

*Medal of Honor Recipients, 1863–1978 (Washington: U.S. Government Printing Office, 1979), p. 664.
†Ibid., p. 655.
‡Ibid., p. 676.

inch shells up the starboard hatch between the third and second decks. With him were two other Sixth Division junior officers, Ensigns I. W. Jeffery and W. F. Cage, also USNR.

Jones was at the foot of the ladder when the bomb ricocheted off the second deck and exploded. Fragments struck him in half a dozen places on his face and body. Jeffery and Cage were also wounded. The two ensigns and a marine private carried Jones up the ladder to the M-Division compartment on the second deck. It was burning, but he refused to be carried any farther.

"Leave me alone!" he said. "I am done for. Get out of here before the magazines go off."*

He died on the second deck of his wounds.

One of the highest honors the navy can bestow is to name a new ship after an officer or enlisted man who has distinguished himself in combat at the cost of his life. Five heroes from the *California* were so honored during World War II: three of the four Medal of Honor winners, Ensign Jeffery, and Seaman Second Class Thomas J. Gary. Jeffery later succumbed to the wounds he received. Gary pulled three or four shipmates from burning compartments and was trying to save others when he fell.

The ships, all commissioned in 1943, were the *Reeves*, DE 156, reclassified as APD 52, Norfolk Navy Yard; the *Herbert C. Jones*, DE 137, Consolidated Steel, Orange, Texas; the *Scott*, DE 214, reclassified as APD 64, Philadelphia Navy Yard; the *Thomas J. Gary*, DE 326, Consolidated, Orange; and the *Ira Jeffery*, DE 63, reclassified as APD 44, Bethlehem Ship Building Company, Hingham, Massachusetts.

If we had our heroes, we also had those responsible for our defeat. The list began with the battleship captains who had not been on board their ships at the opening of hostilities and worked up the chain of command through Rear Admiral P. N. L. Bellinger (commander of all Hawaiian-based naval patrol aircraft) and Pye, Bloch, and Kimmel to CNO Stark, SecNav Knox, and President Roosevelt himself.

In the absence of any explanation for the catastrophe, rumors multiplied like the rabbits that had once abounded on Ford Island. They were circulated with relish by those who delighted in seeing the mighty brought low and were only too glad to kick the prostrate bodies, and with anger by others who were shocked at the carnage and wanted those responsible punished for their blunders. It was more than the death and destruction for which they had to be held accountable; it was the shame of our awful defeat. We had not failed the navy or the country in our duty—our leaders had.

The consequences of their failure reached far beyond the American

*Ibid., p. 586.

casualties of 2,403 dead and 1,178 wounded. The water supply on Ford Island may have been inadequate, but we did have a steady stream of bad news from the Honolulu newspapers by day and from a portable Zenith radio that pulled in broadcasts from San Francisco, Los Angeles, Dallas, and Salt Lake City by night. Massive Japanese air strikes on 8 and 10 December destroyed our air fields and air power in the Philippines, despite the fact that by this time we were sufficiently warned of Japanese intentions. On 10 December, Guam fell. Not to be outdone in grievous underestimation of the enemy, the British sent the battle cruiser *Repulse* and the battleship *Prince of Wales* into hostile seas without air cover. They were promptly sunk off Malaya. On that same day, 11 December, Germany and Italy declared war on the United States. The next day, the Japanese invaded Luzon in the Philippines.

The only light in all those bleak days and blacked-out nights was provided by the 450-man marine-corps garrison that threw back the first enemy assault on Wake Island and sank two Japanese destroyers. They were the first surface ships our forces managed to sink. Even our highly acclaimed submarines had done nothing so far. It was a vivid demonstration of the apparently insoluble problem faced by all navies: the officers who reach command in peacetime are often unsuited to command in time of war.

In the radio dugouts, or when walking along the shores of Ford Island to inspect the fire-blackened remnants of Battleship Row, we talked about all the air-raid drills we had had at Pearl Harbor during the past year. We recalled the practice blackout of the past April, and the alert in October, when the *California* was in Long Beach for "strength through love." The Hawaiian commanders were considering the possibility of air attack then; why not in early December?

It was only necessary to look up to see the airspace over Oahu alive with our planes, despite the fact that we had lost nearly 200 aircraft on 7 December. Now, obviously, we had the long-range reconnaissance that had been so tragically lacking before the Japanese striking force arrived.

We remembered Pye's marine orderly. He had got twenty days in the brig for being asleep at his post outside the admiral's cabin. What, we asked, would be a suitable penalty for our top army and navy commanders, who had been asleep at their posts in the face of the enemy?

At the Ford Island mess hall, our dessert consisted of freshly baked scuttlebutt. Some of it was even true. Word leaked from Fort Shafter, for example, that one of the army's mobile radar stations had picked up the first Japanese attack wave at more than 130 miles and reported it to the information center. The inexperienced and unbriefed second lieutenant who was acting as pursuit officer decided, on his own initiative, that the planes were friendly. We wondered how well he slept nights. But Joseph

L. Lockard, the lowly private in charge of the radar station, became a media celebrity, one of the many unlikely heroes of 7 December. He was promoted to staff sergeant and was soon on his way to officer candidate school in the States.

Most of what we heard was hearsay and speculation of the rankest sort, which serves as a reminder that when men are told nothing, they inevitably invent their own explanations. One of the four Japanese carriers that had launched the air attack was variously reported as sunk or beached somewhere between Oahu and Japan. This peripatetic carrier, the story went, had conducted raids on Midway, Wake, and Guam before meeting our vengeful fliers. Moreover, it was widely thought that all the Japanese flights were the one-way suicide type. (Actually, there were six carriers in the striking force. After recovering all but twenty-nine of their planes, they escaped to the northwest while our befuddled forces were searching to the southwest. "We didn't lay a goddam glove on 'em," was the succinct summary of one radioman.)

Some of the most remarkable scuttlebutt grew out of our inability to accept the fact, despite incontrovertible evidence, that Japan could produce such outstanding pilots and equipment. We were thus victims of our own propaganda. Some of the planes, we were solemnly assured, had even been piloted by blond Germans. They were equipped with RCA radios and Pratt & Whitney engines and were stocked with food, soap, and other necessities.

Every day, it seemed, there was a new submarine scare inside the harbor. Destroyers raced here and there, occasionally dropping a depth charge that shook up both the water and the battleship survivors on Ford Island. All the sightings were false alarms, of course: every slow-drifting tin can became the tip of a Japanese periscope.

The fear that submarines had penetrated our protective nets at the harbor entrance was fueled by reports that two ships had been sighted off Barbers Point unloading sixty-foot midget subs. When I heard this hysterical rumor, I had to laugh.

"The navy has got things well in hand," I told an appreciative audience of radiomen. "We've sunk the forty transports that were off Barbers Point. Now we're down to two sub tenders!"

Through it all, the navy enlisted man exercised his right to "chip his gums," curse his officers, and in the darkest moments, discover the saving grace of humor, as I had done, as Americans always have. Things were so bad at Pearl Harbor, we told each other, that even the chiefs were working.

By now a rough pontoon bridge—planks secured to empty gas drums—had been run from Ford Island to the *California*. As the ship continued to settle into the mud and slide away toward the channel, the numerous manila lines and wire hawsers that secured her to the quays had begun to snap with sounds that resembled the firing of high-powered rifles. Trying to get aboard involved some risk of decapitation or disembowelment. The bridge appreciably reduced this hazard. Finally, the ship's anchor chains were rigged to the forward and after quays, and the *California* reached an uneasy equilibrium.

With the danger of capsizing past, plans were made to shift the emergency radio center from our dugouts to the flag conning tower. Sent to flag conn on an errand, I encountered Captain Bunkley, bent over and inspecting near-miss damage to the starboard bow. I hadn't seen him since well before the attack. Dirty and unshaven as I was, I gave him a smart salute. Straightening, he returned it briskly.

The captain looked older. He must have been badly shaken by the sinking of his ship and his career. I remembered hearing that he had been observed the afternoon of the attack, sitting on a bitt with his head in his hands. "A sad specimen," in the words of one of the junior officers.

I felt a little sorry for him. I could understand why captains sometimes elected to go down with their ships or, if their performance did not meet the highest standards of the naval service, committed suicide. That is what Kapitan zur See Hans Langsdorff of the *Admiral Graf Spee* did after his pocket battleship had been defeated by the British and he had been forced to scuttle her at Montevideo.

I left the *California* for the last time in mid-December. This time I didn't look back as the launch departed the ferry landing and headed cross-channel for the sub base. I wanted to remember my ship the way she had been, not the way she was now.

When I had left her on that liberty of 5 December 1941—less than two weeks before in time but an eon in experience—she was the flagship of the Battle Force. Then she was a thing of strange, robust beauty, manned by proud and profane professionals. Now she looked in truth more like a monstrous and cluttered barge than a battleship. She was low in the water and still had an awkward list of six degrees. Her forecastle was awash at high tide. The secondary battery had been removed, and nine of her twelve fourteen-inch guns soon would be. Preparations were being made to cut down her mainmast, where the angel of death had brushed me with barbed wings. Divers were below, patching the torpedo holes. Oil was still oozing from her wing tanks. Her once-spotless teak quarterdeck was under water to the top of number four turret. She was coated with soot and smeared with fuel-oil residue. And that was only what one could see. Below decks, she was indescribably filthy. Many bodies were still trapped in the mess.

I certainly did not want to remember her that way.

I departed from the haunted island with all my worldly possessions: the

In this photo of the *California*, taken some time after the author left his "dugout duty" in the emergency radio station on Ford Island, the mainmast has been cut down, nine of her twelve fourteen-inch guns have been removed, along with the secondary battery, and cofferdams are in place around her forecastle and quarter-deck. The bow is just breaking water, the result of extensive dewatering. (Official U.S. Navy photograph.)

clothes I was wearing, my billfold, a half-empty package of cigarettes, a helmet, and a gas mask. Like all the radiomen who had stood watches with me and slept in the dugouts on piles of life jackets, I hadn't shaved or showered for more than a week. The thing most valuable to me, next to my good-luck photo, I had had to leave behind for my replacement: the Springfield.

At the sub base solicitous CinCPac radiomen once again took me in tow. I was provided with a toilet kit, razor, and soap. Again I luxuriated in a long hot shower, this time without the kerosene pre-scrub. I was able to draw some undress whites, dungarees, and accessories, with the promise of a full seabag later. However, no personal items would be replaced. In my locker, awash in the filth of the *California's* third deck, were my *Bluejackets' Manual, The Collected Works of Shakespeare, The Rubaiyat of Omar Khayyam,* letters, snapshots, and a few phonograph records I had collected, including Glenn Miller and the Dorseys.

Having never had the chance to become attached to material things, I shrugged. Most of them could be replaced, now that I was going to draw my pay. The things I truly valued could never be replaced.

No one burdened me with watches. I got some decent meals, joined friends at the band concerts in front of the ad building, and got mildly high on green ten-cent beer. And, finally, I had a chance to get letters off to my foster-parents, father, and great-uncle.

From Ford Island, I had been able to send nothing but stark survivors' postcards. The message side consisted of typeset lines that one could leave in or scratch out, as appropriate, to indicate whether one was well, admitted to the hospital wounded or sick, and whether the condition was serious.

I learned later that I had been reported missing in action. Fortunately, the survivors' cards arrived shortly afterward. Apparently, the ship's chaplains never got the word that I was safe in the radio gang dugouts only a few hundred yards from hangars 37 and 65, where most of the *California* crew were quartered. (Jack Rossnick, who was pressed into duty at the Makalapa Transmitter Station by Flag Chief Reinhardt, was also reported missing in action.)

With a couple of buddies, I made a trip to Hospital Point to visit C-D Division casualties. We walked through long wards filled with burned and wounded sailors. It was not so hard to pass those who were maimed: their missing or mangled limbs didn't show beneath the white sheets and coverlets. But the burn cases shocked me profoundly. Many of those wearing the T-shirt-and-shorts uniform of the day had been burned on every unprotected part: arms, legs, faces. Their skin was as black and crusted as if it had been exposed to blow torches. Long strips of flesh were hanging loose. The patients were being treated with sulfa powder, which added to their grotesque appearance. I didn't see how any of them could ever look normal again. (Most of them did eventually recover.)

In one of the wards, we found our radio gang mates. A handsome young striker showed me his leg. It was horribly gouged by shrapnel and withered to half its normal size. Much to our astonishment, his mother was with him. Somehow, she had wangled passage to Honolulu on the Pan Am Clipper. A youthful-looking forty, with presence and open charm, she was as delighted to see us as she was furious at the navy for permitting this to happen to her son. I would not have been surprised had she announced plans to picket Kimmel's office. She asked if there was any chance the captain would visit his wounded. We said that we doubted it very much. She looked disappointed and angry. I was not sure I wished even Captain

Bunkley so much ill fortune as an encounter with this formidable American mother.

After one or two more visits, I had to stop going to the hospital. Maintaining false good cheer in the presence of so much pain, so many shattered bodies and lives, seemed hypocritical. Moreover, I was forcibly reminded of what could happen to anyone in combat, a thought I could not afford to dwell on. At the price of a certain insensitivity, I had to harden my resolution. I knew I would be going to sea again and that there would be other engagements. The "day of infamy" had left me with substantial doubts about the abilities of my leaders. Would I be given an opportunity to repay the enemy for the deaths of Johnson, Reeves, Gilbert, Beal, and all the others—or would I be led into another American bloodbath?

On 17 December, Kimmel was charged with "dereliction of duty" and summarily relieved of his command. On this count there were few dissenters among the radiomen. Admiral Pye was named commander in chief *pro tem* while a new CinCPac was sent out from Washington. (Admiral Chester W. Nimitz relieved Pye on 31 December 1941.)

By now, the *California* crewmen had heard that Pye had lent his authority to Captain Bunkley's abandon-ship order at 1000. We also knew that the black gang had been ready to get under way since about 0915, using four boilers. When the fire rooms and engine rooms were secured forty-five minutes later, the ship's fate was sealed. We wondered whether Pye was capable of making tough decisions under combat conditions. Nothing could be done but to give him the benefit of these doubts in the hope that he would make every effort to restore luster to his tarnished reputation.

He was put to the test immediately. Wake Island was in imminent danger of falling. Without reinforcements, it could not hold out much longer. A Wake relief expedition built around the *Lexington*, *Saratoga*, and *Enterprise* had been organized and dispatched by Kimmel—his last official act. Now its employment was in the hands of Pye.

On 21 December, Fisher and I went up to CinCPac headquarters on the second deck of the ad building. The atmosphere was informal and hectic. The grave faces of officers and men told us that great events were afoot involving Wake Island. We saw Pye emerge from his office, trailed by two or three of his aides, and rush down the hall to what we assumed was the situation room. The only time I had seen the admiral move that fast was when he boarded the *California* midway through the attack. He seemed a little distraught, heavily burdened by his awesome responsibilities.

Fisher and I exchanged glances. I knew he was thinking the same thing I was: "Situation normal, all fouled up." It had been that way ever since the *Ward*'s contact report had been bucked up the chain of command and passed back down again for verification. Two country boys from Placerville had been given the rare privilege of seeing some of their leaders at close range. They did not like what they saw.

Faced with his most important command decision, Pye faltered, as Bunkley had done. He vacillated, issued conflicting orders, and at last called back the relief expedition. Wake Island fell.

On the flag bridge of the *Saratoga*, the talk was "mutinous." Marine aviators cursed and even cried. In the *San Francisco*, some of Rear Admiral Frank Jack Fletcher's staff urged him to ignore the order. When Admiral Joseph M. Reeves, a former CinCUS and a member of the Roberts Commission investigating the attack, learned what had happened, he reacted in words that were reminiscent of my late friend Johnson: "By Gad! I used to say a man had to be both a fighter and know how to fight. Now all I want is a man who fights."*

One day in late December, Lieutenant Sugg found me in the CinCPac radiomen's quarters. My undress whites were pressed. I was clean-shaven and my hair was regulation length. I looked like a battleship sailor again.

"Ah, Ted," he said genially. "It's good to see you. I've got good news. You've been selected for the ComBatFor flag."

That meant, presumably, that I would be going to sea with Admiral Pye again. If this qualified as good news, so did the British surrender of Hong Kong on Christmas Day.

"I'll be going with you," Sugg said reassuringly. "Meanwhile, how would you like a little detached duty with the army signal corps?"

"The army signal corps, sir?"

Sugg explained that the *California*'s CXAM radar had been removed and set up on a spur of the Koolau Range, behind Fort Shafter, where it would be operated in conjunction with one of the army's long-range search radars.† The *California*'s two regular CXAM operators, Vernon Luettinger and F. C. Tarlton, would be going along.

"We need a third operator so we can maintain a twenty-four-hour watch," Sugg said. "I believe you're just the man for this assignment."

"But, sir, I've never operated the CXAM before. I've never even seen it." The CXAM was one of the navy's top-secret defensive weapons. No one except Commander Bernstein, Ensign J. N. Renfro (division officer of C-C and C-K divisions), and the two enlisted operators had been allowed inside

*Morison, Samuel Eliot, *History of United States Navy Operations in World War II* (Boston: Little, Brown and Company, 1967), vol. 3, pp. 252, 254.

†CXAM Serial Number One, the first operational radar in the fleet, had a short and most dramatic history. It was removed from Fort Shafter and installed on the *Hornet* (CV 8) in the summer of 1942. It went down with the *Hornet* on 27 October 1942, during the Battle of the Santa Cruz Islands—twenty-six months to the day after its installation on the *California* had been completed at Puget Sound Navy Yard (27 August 1940).

the locked compartment off the signal bridge. The only part of the radar I
had seen was the "bedspring" antenna mounted atop the superstructure.

"Well, you'd never copied NPM Fox in a dugout alongside a sunken
battleship, either," my radio officer replied in jocular fashion. "With your
brains, Ted, you'll pick it up in no time."

No wonder Lieutenant Sugg had quickly become a big executive with
NBC Radio Network, I thought. His combination of cajolery, flattery, and
the subtle hint of iron purpose was almost impossible to resist. What could
one say?

"Aye aye, sir!"

Compared with the Ford Island pits, Fort Shafter was a rest center.
Luettinger, Tarlton, and I lived in a tent with a board floor and slept on
army cots. Nearby was a wooden latrine with running hot and cold water.
We were given metal mess kits and took our meals with the signal corps
soldiers in a roomy mess hall. The food was not up to navy standards,
leaning heavily toward canned peas, mashed potatoes, cold "horsecock"
(Spam), and the interservice "SOS". But, then, the fare at the Ford Island
mess hall was not even up to Georgia chain-gang standards.

An army truck took Luettinger and me up a precipitous dirt road to the
radar station. The CXAM was housed in a small wooden shack about fifty
feet away from the truck-mounted van that contained the army's SCR-270
radar. The roof of the shack supported the seventeen-square-foot, wire-
curtain-type antenna I remembered. Inside, I had my first look at the
electronic marvel that was about to revolutionize naval warfare.

The CXAM occupied two side-by-side consoles, each nineteen inches
wide by about six feet high, in the usual black crackle finish. The transmitter, with its switches, meters, and antenna-control handwheel, was on the
right. At eye level on the receiver was a green-phosphor cathode-ray tube
seven inches in diameter. Outlined in grassy green on the CRT was a
rectangular display, called a folded-A presentation. It was scaled in miles,
from zero to fifty, like a ruler.

Suddenly, a large blip rose above the green line at about forty-five miles
and then decayed as the antenna continued sweeping the horizon at four
revolutions per minute.

"We have a bogey," my shipmate said. He was seated on a high stool. I
was peering over his shoulder in the darkened shack.

When the antenna made its next rotation, the target showed up again. It
was perceptibly closer. A plane was coming in, just as the Japanese planes
had come in that Sunday morning. They, too, had announced their arrival
on a marvelous cathode-ray tube like this one.

"Lute" turned the handwheel to its neutral position, stopping the
antenna. Manipulating the wheel with the dexterity of long practice, he
rocked the antenna back and forth until the pip was at its maximum

amplitude above the display line. Referring to a synchro repeater dial, he read the true bearing of the target, which continued to move closer. He pushed the button of his headset.

"Reporting a bogey at four-two miles, bearing two-three-five true," Lute said to a plotter at interceptor control, Fort Shafter. It was the same information center that had shrugged off the report of the first wave of Japanese planes. They were not shrugging off radar reports now.

Lute put the antenna back onto automatic rotation. "Probably a patrol plane coming back to Ford Island," he said.

I was fascinated. I could hardly wait to operate this fantastic electronic spy that could reach out day and night, in rain or fog, to detect the presence of unseen aircraft. Properly employed, radar would have given us perhaps a thirty-minute warning, time enough to button up the open voids, set watertight Condition Zed, and go to battle stations. Then we would have had a fighting chance—and that was all we asked for. I stood a couple of

The author at the entrance to his tent while on detached radar duty at Fort Shafter, Honolulu, March 1942. Note the mixed army-navy uniform; he had yet to get a full seabag following the sinking of the *California*.

instruction watches with the ever-polite and patient Luettinger. Then I was on my own.

On board the *California*, in peacetime, I couldn't get near the CXAM. Now, with a war going on and the stakes infinitely higher, I was one of three men trusted to operate it. That, I was sure, had been Sugg's decision: he had shown great confidence in me, and I was grateful for it. In turn, he had asked a good deal of me, but no more than was my duty. I wondered why more officers didn't try Sugg's system of dealing with enlisted men. With that kind of treatment, there was scarcely a limit to what men could and would do.

For a while, I lived the life of a rather privileged sailor in an army camp. The discipline was lax by navy standards, and the soldiers treated me well. From the top of the arête, I enjoyed a dramatic view of the approaches to Pearl Harbor.

I watched the old picket destroyers in their purposeful meanders and saw the combat ships come and go. In mid-January, the *Saratoga* slipped in with a slight limp and what almost seemed a hangdog expression: she had been torpedoed, for the first but not the last time, by a Japanese submarine 500 miles southwest of Hawaii. All day long, planes were overhead on the patrol missions Kimmel had said he could not maintain because they interfered with training and maintenance schedules. The surprise attack had had a remarkable effect on the ordering of priorities.

Placidly, the Pacific rolled in on the Oahu shores, heedless of man's violent enterprises. It was hard to imagine that across the serene horizon we and the enemy were engaged in life-and-death struggles, and that with every one of them sorrow came to homes in the United States and Japan alike.

With nightfall, the atmosphere changed. I sweated in the airless confines of the radar hut. The luminescent green scope provided the only light. We were still fearful of sabotage and even more wary of the trigger-happy army infantry troops who guarded the two radar installations.

The approach of dawn brought maximum danger of air attack. I tracked our planes out until they disappeared from the screen, watching the moving blips with satisfaction. Then I tracked them back, careful to report every weak or strong bogey, lest one of them prove to be the enemy.

I needed this interval of duty with little hazard, away from the stinking abattoir that was Pearl Harbor. Wearing a sweatshirt, I did roadwork around the compound with Corporal Stanley Kearns, a former army middleweight boxing champion. At first, I was out of shape and would start lagging behind after a couple of miles.

"C'mon, Seaweed," Kearns would shout back, increasing his speed. "Pick 'em up and lay 'em down!"

Around the fort I wore navy dungaree pants with a webbed army belt and a tan GI shirt; when I went to Honolulu with Kearns and other soldier friends, I wore my undress whites. I was surprised to discover that the soldiers who had been stationed at Shafter for some time fared better with the residents of Honolulu than the sailors. My new buddies took me to Hawaiian homes and introduced me to dark full-figured Kanaka girls. I drank Five Islands whiskey with them and their large, laughing mothers.

One came back. The author's two best friends, photographed at Waikiki in November 1941. Floyd C. Fisher (*left*) survived the Japanese attack to join the author on the Commander Battle Force flagship *Pennsylvania*. Melvin G. Johnson joined Thomas J. Reeves, Tom Gilbert, Joe Ross, and other heroic dead at the National Memorial Cemetery of the Pacific in Honolulu.

Again to my surprise, I found the Hawaiians to be truly accepting, hospitable people.

But this pleasing insight came a little late. There was a war on, and I was a member of the flag complement of what was left of the Battle Force: the *Pennsylvania, Maryland, Tennessee,* and *Colorado,* shortly to be augmented by the return of the *New Mexico* and *Mississippi* from the Atlantic Fleet.

All too soon, my relief arrived. I said reluctant good-bys to Luettinger, Tarlton, Kearns, and the other soldiers.

Lieutenant Sugg had not forgotten me.

I stood on the midships hangar deck of the heavy cruiser *Louisville* (CA 28) as she stood out of Pearl Harbor on Sunday, 29 March 1942. Her mission was to deliver Admiral Pye and his staff and flag allowance to San Francisco, where we would board the *New Mexico.* (We reported to the *New Mexico* on 3 April. Admiral Pye soon changed him mind. We repacked our seabags and shifted to the *Pennsylvania.* "No wonder Pye lost Wake Island," Red Goff griped. "He can't even make up his mind which flagship he wants to go down in!")

After nearly four months ashore, it seemed strange to stand on the decks of a combat ship again. To a battleship sailor, the *Louisville* looked frail and narrow of beam. One of the *Northampton* "treaty" class of 1924, she had a standard tonnage of only 9,050 and was very weakly protected. I could vividly imagine what one of the murderous Japanese torpedoes would do to this eggshell cruiser. But, I reassured myself, her 107,000-hp geared-turbine drive turned up nearly thirty-three knots, which should make us a difficult target for any lurking Japanese submarines.

On board with me were some twenty of the *California* radiomen: Fisher, Goff, Raphalowitz, Mazeau, and Joe Goveia among them. I was pleased that two other friends, Tom Moore and J. K. ("Jawbones") Madden, had also been elevated to flag status. Reinhardt was chief in charge, inheriting Reeves's position. Another chief named D. L. Hyde was along to run the transmitter room, as he had in the *California.* Sugg had become our radio officer. I certainly had no reason to dissent from the near-unanimous approval. Promoted without the necessity of taking an examination, I was sporting the rating badge of a second-class petty officer. Commander H. O. Hansen, alas, was still flag communication officer. More than ever, I had no intention of going to flag radio.

Missing were a number of our "hotshots": they had been selected for duty with CinCPac, who badly needed their special operating and technical skills. They included Dave Rosenberg, Jack Rossnick, Joel Bachner, and Dick Stammerjohn. Even as we jokingly accused them of "brown-nosing" themselves into safe shore duty, we were suitably envious. My sense of survival had to struggle against my sense of duty. I knew the war would be won at sea, in battle with the Imperial Japanese Navy, and not at NPM Radio, no matter how important fleet communications were. My comrades of the ComBatFor flag and I were among those who would have to win it. I wished I had more confidence in my leaders; but if my late friend Johnson had been right, and I was sure he had been, a new breed of tough, aggressive commanders would be coming along soon to replace the blunderers.

Missing and beyond the call of duty were Chief Reeves, M. G. Johnson, Tom Gilbert, A. Q. Beal, Joe Ross, and Al Rosenthal. The other radiomen who had been killed in action were Donald C. V. Larsen, Frank W. Royse, and Frank P. Treanor, all radiomen third class. One of our strikers, Douglas D. Drysdale, was missing and presumed lost. Several other radiomen were still in the hospital. I thought about all of them as I stood on the hangar deck.

I thought especially of the radioman who had gone back to sleep that Sunday morning in the belief that general quarters was "another goddam drill." He was among the dead. Should I have stopped to drag him from his bunk? I had decided that my ship came first. It was one of those spot decisions that one has no time to consider but a lifetime to think about. His loss was an object lesson I did not need. Because he had a hangover and because he had not learned to obey orders instantly, he had died in vain.

Commander Bernstein was not with us. He had been transferred to the Bureau of Ships, where he would serve with distinction during the war as head of shipboard radar design, and later as head of electronic installation and maintenance.

Captain Bunkley had been transferred to the Fourteenth Naval District for temporary duty. He later served as supervisor of New York Harbor from mid-1942 through January 1946.

Our philosopher second class had been shipped off to the heavy cruisers, as had Jerry Litwak. I had last seen the former while on dugout duty. He had been far too shaken to give me the expected "I told you so!" All he could do was repeat, over and over, "Jesus Christ! I never thought it would be this bad!"

Hundreds of other crewmen had been transferred to the cruisers *Portland* (CA 33), *Indianapolis* (CA 35), *Astoria* (CA 34), and *Chicago* (CA 29). In less than a year, *California* alumni in the latter two heavy cruisers would be abandoning ship again—this time off two South Pacific islands we had never heard of, Savo and Rennell.

As the *Louisville* made the big turn to port around Hospital Point, I looked back at Battleship Row. I didn't want to, but I couldn't help myself.

The *Maryland* and *Tennessee* had each taken two bombs; our sister ship also had considerable fire damage and warped stern plates resulting from the explosion of the *Arizona* close astern. Both had been cleared from their

Tugs slowly move the sodden hulk of the *California* toward dry dock at Pearl Harbor. Following a complete modernization at Puget Sound Navy Yard, she rejoined the Pacific Fleet. She was decommissioned and placed in the Reserve Fleet at Philadelphia on 14 February 1947. On 15 January 1959 she was sold for scrapping. (Official U.S. Navy photograph.)

quays, repaired, and were en route to California by late December. The *Pennsylvania* had taken one bomb and had fire and fragmentation damage to her bow; she, too, had departed for San Francisco before the end of 1941.

But the *California* was still at Berth F-3. After months of dirty, disagreeable, and often dangerous salvage work, she had been refloated on an even keel and would very soon be moved across channel into Dry Dock Number Two. With all her guns (except for number four turret) removed, her mainmast chopped down and wooden cofferdams in place around her forecastle and quarterdeck, she was little more than a sodden hulk of a battleship.

I felt sick at heart as I looked at her. Perhaps her plates would be patched and her turbines, boilers, and electric motors reconditioned so that she could reach Puget Sound under her own power. Perhaps the battleship experts there would repair and modernize her so she could rejoin the fleet. She might even fight again—this time, I hoped, at battle, not slaughter, stations. But she would not look like the *California* I knew. She would have a different role and purpose, a far more modest one. No more would her decks echo with the measured tread of three- and four-star admirals. The *California* I knew, the flagship of the Battle Force, was gone.

Astern of Battleship Forty-four, in a crooked line, were the *Oklahoma*, still upside down and looking, as more than one sailor had remarked, like a huge, broaching gray whale; the *West Virginia*, still embedded deep in the mud, water lapping at her boat deck; and the *Arizona*, her forestructure still crazily awry, a somber Davy Jones locker for 1,104 American sailors. (Of these three, only the *West Virginia* would ever be reclaimed; and she would not depart Puget Sound for the war until mid-1944, when the issue had already been decided.)

Looking aft from the hangar deck of the *Louisville*, I knew I had been an eyewitness at the death of the battleship navy. The former queen of battles now was reduced to the status of mere lady-in-waiting on the newly crowned, but decidedly unmajestic, flattop.

Never again would the mighty battlewagons of the Pacific Fleet steam off the California coast in line of battle ahead, their turrets trained out against the marauding forces of Enemy Orange. Never again would the fourteen- and sixteen-inch rifles speak in concert, splitting the heavens with earthly thunder for the benefit of the target raft, the official observer, and the newsreel camera. Now, if the battleships spoke at all, it would be in earnest; but the enemy was to the west and they had been, temporarily at least, withdrawn to the east. Our humiliation was complete.

Never again would the once-revered battleships anchor in arrow-straight column in the lee of the San Pedro breakwater. And never again would the fleet landings at Long Beach and San Pedro come alive with thousands of battleship sailors, bringing good news to bar and bistro and

The *California* arrives at Puget Sound Navy Yard on 20 October 1942 for refitting and modernization. Permanent structural repairs to most of the torpedo and bomb damage had been performed in dry dock at Pearl Harbor, and the main battery had been reinstalled. Many bodies had been removed from the second and third decks. But to any battleship sailor she was still a sad sight. The entire mainmast had been removed. The cage foremast and battle top were gone, too, replaced by a prosaic stick mast. Missing also were the broadside guns, the 3-inch and 5-inch antiaircraft batteries, the catapults and their planes, and most of the boats: everything that had made the *California* a graceful and formidable capital ship of the line. (Photo courtesy of Puget Sound Naval Shipyard.)

bordello; to tailor and pawnbroker and credit jeweler; to the sailor's girl in furnished room or hotel or apartment.

*The fleet's in!* were words that would echo now only in the chambers of the memory.

And what would the Fourth of July be, when it did not find the pale gray battleships with the symmetrical twin masts anchored off San Diego and San Clemente and Monterey and Santa Cruz and San Francisco and Portland and Tacoma and Bellingham and Port Angeles? To the citizens, the battleship was a symbol of the nation's power, bringing them a thrill of patriotism. So long as those great, beamy ships of the line were out there, they knew that the nation's continued freedom was ensured.

And what about those impressionistic lads, to whom the lean, clean, aloof battleship sailors brought the romance of the sea, as that young petty officer had for me on a certain Independence Day at Santa Cruz? I did not think sailors of the new queen of battles would cut quite the valiant figure that their predecessors had.

There was pride, understated but ever present, in being a sailor in the battleship navy. I had felt it every time I stepped off that holystoned teak quarterdeck and saluted the flag from the top of the accommodation ladder; every time a civilian asked me what ship I was on and I replied, a little smugly, "The *California*"; every time I admired my ship from a motor launch, my eyes running from her rakish clipper bow past the turrets and the piled-up superstructure and the tall cage masts to the national ensign rippling over the water astern. And especially had I felt it when the rising arpeggio of the bugle sent me double-timing to my battle station.

Now the battleship navy was no more, after forty-five years of proud service as America's first (and often only) line of defense. It had been blasted away in two hours of concentrated attack by torpedo and bomb. The battleship had proved vulnerable to a deadly new weapon: carrier aircraft. She had had no chance to defend herself in honorable combat because the enemy had given no overt warning, and because our leaders had proven unworthy of their trust. For this most of them would pay a high price—in disgrace, in loss of credibility, and worst of all, in being denied a chance to redeem themselves. But not so high a price as 2,000 American sailors had paid. When great men blunder, they count their losses in pride and reputation and glory. The underlings count their losses in blood.

The battleship navy was gone as an overwhelming physical force; but I knew that some of its intangible legacies would remain. What would live for others to emulate and for history to admire were the qualities of men like Reeves and Pharris and Scott and Jones and Gary and Beal. To these men, defeat was unendurable, failure was not acceptable.

This true martial spirit was not exclusively a battleship navy virtue—but it surely was part of the battleship mystique. If such a spirit were ever to

The *California* under way in camouflage paint in January 1944, following renovation at Puget Sound Navy Yard. She now has a solid tower superstructure and a stumpy fire-control platform where her mainmast had been. The mixed secondary battery has been replaced by sixteen 5-inch 38-caliber dual-purpose guns in shielded twin mounts, along with new fire-control directors and new air-search and fire-control radar equipment. In place of the few antique 50-caliber machine guns that had proved so ineffectual at Pearl Harbor are forty-eight 20-mm Swiss Oerlikons in single mounts and forty of the new 40-mm Swedish Bofors in quadruple mounts.

At Surigao Strait on 25 October 1944, the *California* poured sixty-three 14-inch projectiles into the Japanese battleship *Yamashiro* in the last battle-line engagement in history. After two decades of preserving the peace, one crushing defeat, and one ignoble sinking at quayside, the *California* had risen from the mud of Pearl Harbor to fulfill her destiny.

disappear, the loss to the nation would be far more grievous than the destruction of any number of battleships. Or carriers.

For the fourteen months preceding Pearl Harbor, I had been part of the battleship navy. I and my shipmates of the *California* had absorbed some of its pride and smartness and mystique.

These things had served us well on 7 December 1941. We had stood our ground, and we had done our duty. Some of us had done more. But only a very few had done less.

We might have been defeated, but we were not broken by defeat. Ahead, there was a war to win—and we knew we would win it.

I would always be proud that my shipmates and I had measured up.

I would always feel privileged to have been a sailor in the battleship navy.